EUREKA MATH² CALIFORNIA

MU01716756

LEARN

A Story of Units®

Grade K ▸ Part–Part–Total

GREAT MINDS™

Great Minds® is the creator of *Eureka Math®*, *Eureka Math²®*, *Wit & Wisdom®*, *Arts & Letters™*, and *PhD Science®*.

Published by Great Minds PBC
greatminds.org

Printed in the USA
B-Print

1 2 3 4 5 6 7 8 9 10 QDG 30 29 28 27 26 - 19940

ISBN 979-8-89417-318-4

What's Included

Welcome to *Learn*, students' companion text to the instruction in *Teach*. This resource contains all the pages that students need, including in-class work pages, glossaries, and opportunities for practice.

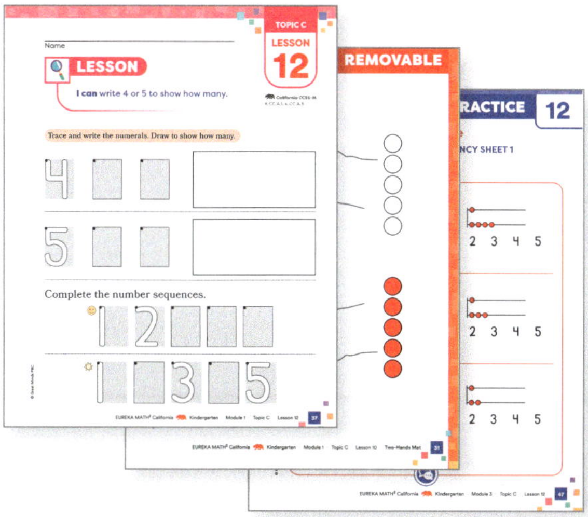

Lesson pages needed during the lesson are labeled by lesson number and include learning outcomes. Pages that include an **orange bar** are meant to be separated from the book and used during class.

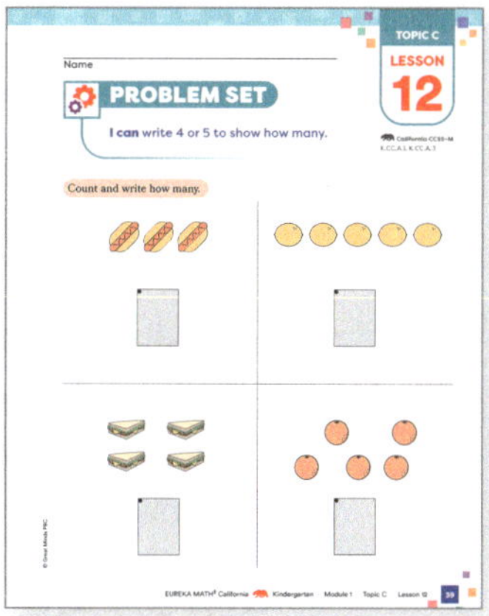

The **Problem Set**, which follows the new learning of the lesson, offers daily independent practice with problems organized from simple to complex for easy differentiation.

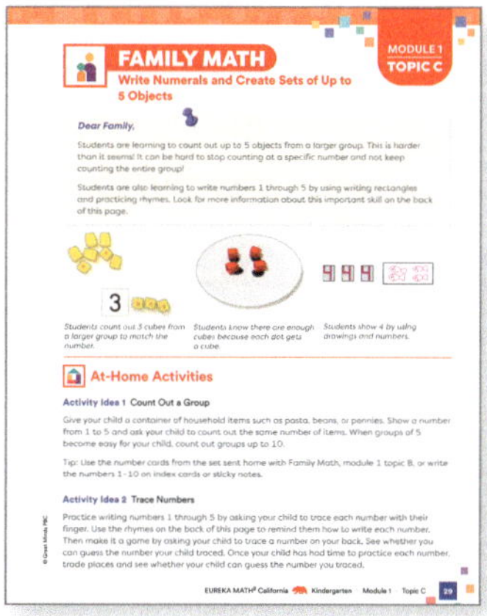

Keep families connected to math in the classroom with **Family Math** letters. Each letter is provided by topic and includes the major concepts and vocabulary. Each letter also includes simple and practical at-home activities to extend the learning and help students see mathematics in their world.

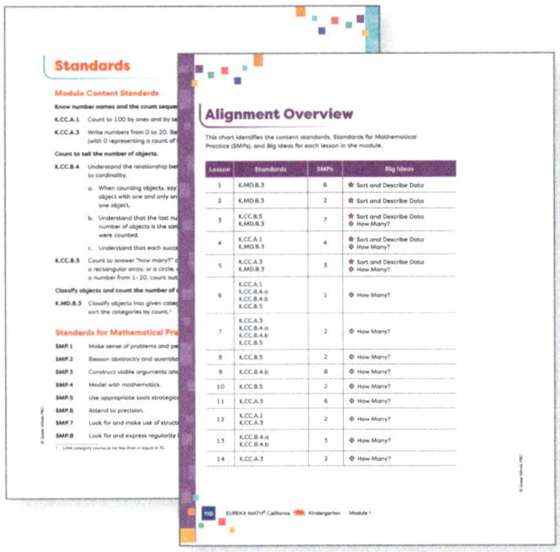

Standards and **Alignment Overview** connect each lesson to embedded content standards, Standards for Mathematical Practice, and big ideas in mathematics.

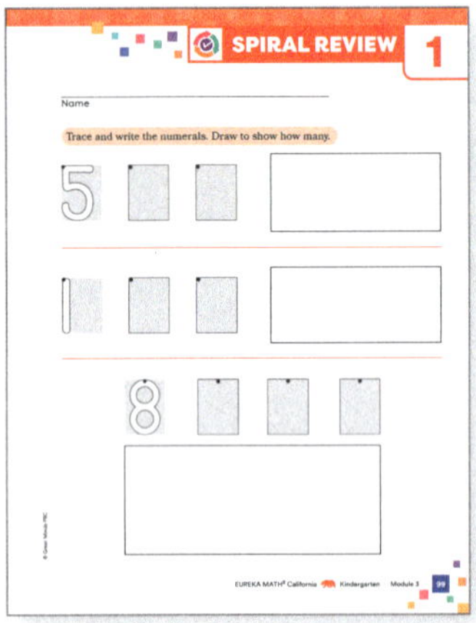

Spiral Review reinforces skills and concepts through distributed and interleaved practice.

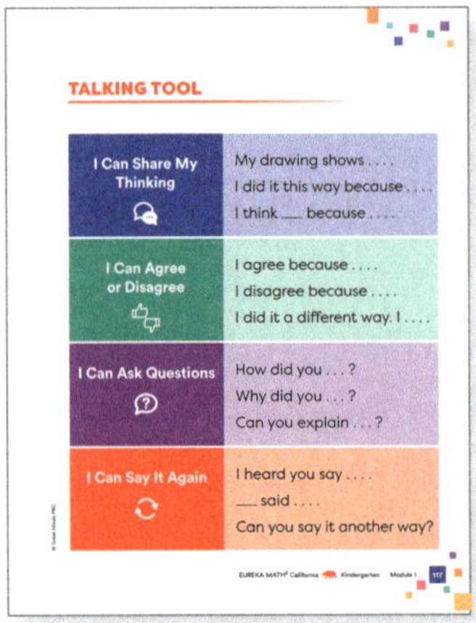

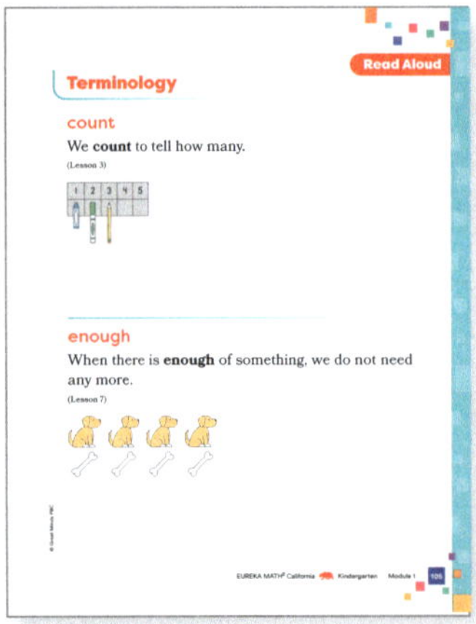

Mathematical language is critical to helping students understand math concepts and describe mathematics in their world. The **Talking Tool** and **Terminology** resource can help students develop fluency with mathematical language and provide supports for discourse.

Contents

Addition and Subtraction

FAMILY MATH

Represent Addition

Dear Family,

Students solve a variety of addition problems and use number sentences to represent their thinking. They make an important transition in how they write and say number sentences. Before, they used common language to describe part-total relationships, "7 and 2 make 9" or "9 is 7 and 2." Now they will learn to read number sentences using mathematical language, "7 plus 2 equals 9" and "9 equals 7 plus 2." Students also explore many different ways to represent addition by using concrete objects and drawings.

Words We Are Learning

Addition

Plus

Equals

[7] and [2] make [9].

[7] [+] [2] [=] [9]

Mathematicians use special symbols to write a number sentence.

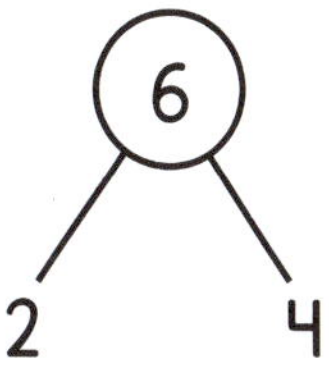

6 is 2 and 4.

6 = 2 + 4

At-Home Activities

Activity Idea 1 Dot Cards

Cut out and show your child a dot card for 3–5 seconds. Ask your child to say how many dots they see. Discuss the different ways to group or see the dots to determine the total. Your child might point to explain, such as, "I saw 2 on that side." (*Points to 2 dots.*) "I saw 4 on that side." (*Points to 4 dots*). "2 and 4 is 6." Invite your child to draw a number bond or number sentence to match the way they see the dots.

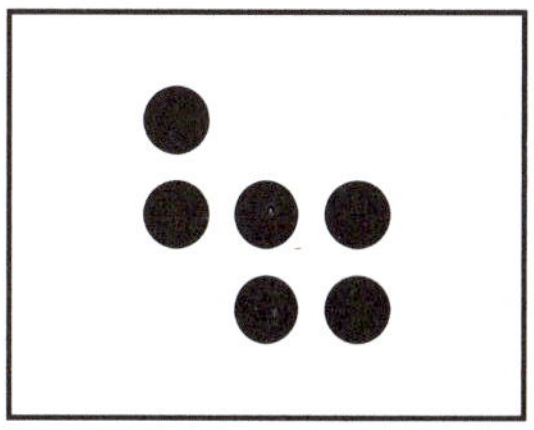

Activity Idea 2 Imagination Stories

Invite your child to gather a group of 10 or fewer small objects. Model telling an addition story and ask your child to show the story using the objects. For example, "There are 5 bunnies playing in the grass." Pause to let your child show you the 5 objects that represent the bunnies. "2 more bunnies come to play." Pause to let your child model 2 more objects joining the group. "How many bunnies are there now?" Give your child time to count to find the total.

Repeat the story with new numbers or let your child create their own story. Prompt them to write a number bond or number sentence to match their story.

Writing Number Sentences

Students learn that when we add, we put the parts or groups together. The number bond helps organize information into parts and a total. The formal introduction to addition begins with stories. The context of a story helps a young student understand what the numbers mean and how to solve the problem. For example, "There were 5 children riding a rollercoaster. 3 children get on. How many children are on the rollercoaster now?"

Students know how to draw a picture and number bond to match a story. Now they can take it a step further to write a matching number sentence. A number sentence is another way to tell the story. They learn that mathematicians would tell the story or read the number sentence as, "5 plus 3 equals 8." A storyteller might read it as, "There are 5 children riding a rollercoaster. 3 more children get on. There are 8 children on the rollercoaster now."

As story contexts change, students learn that number sentences can adapt. In some cases, it makes sense to write the total first. Consider sharing the following story: "There are 5 teddy bears in the picture. I see that some have bow ties and others do not. What number sentences can we write to match the picture?" Students can begin with the total number of objects and then sort to create the parts. They discover that each total can be written as multiple combinations of parts.

5 is ☐ and ☐

5 = ☐ + ☐

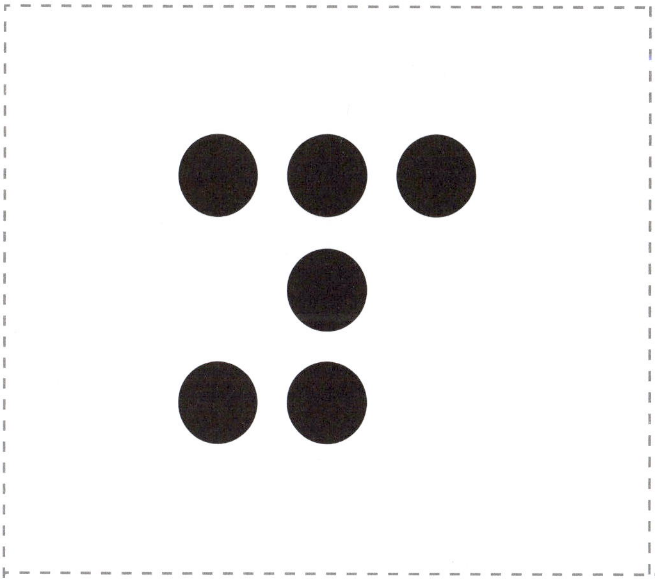
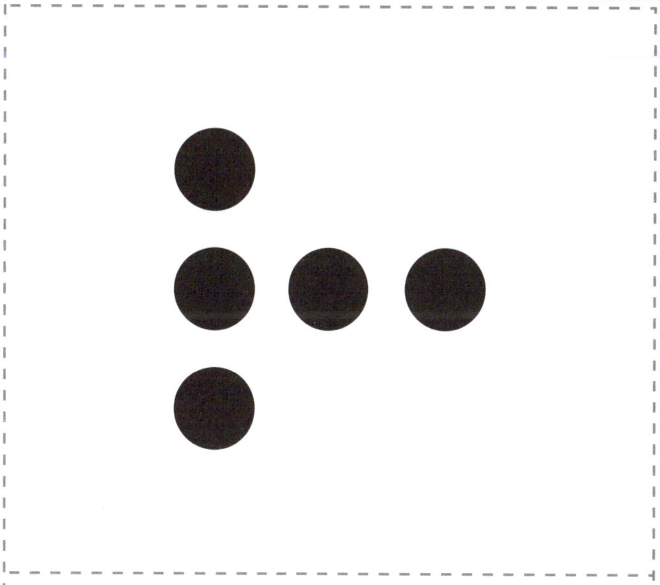
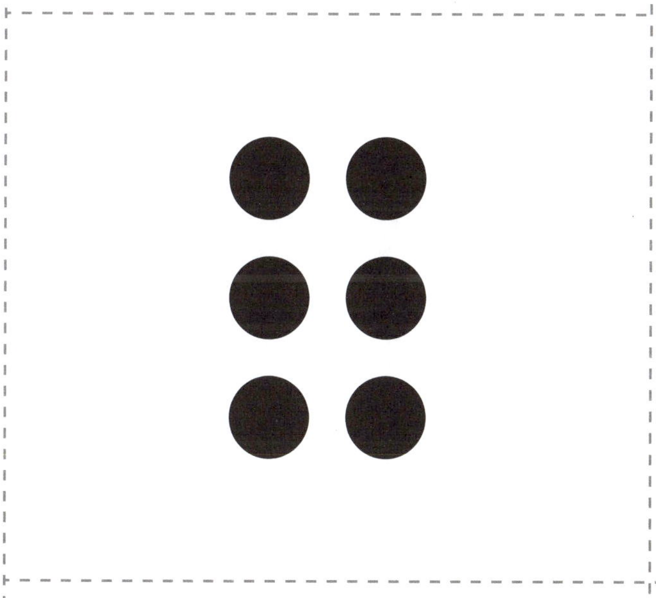
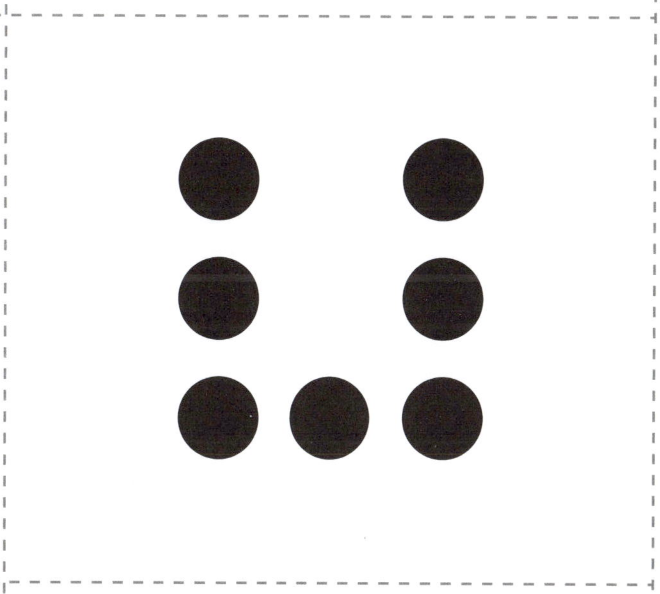
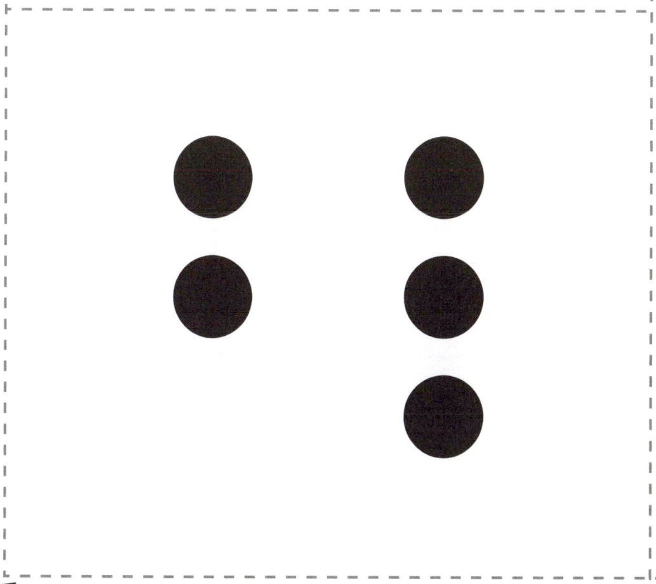
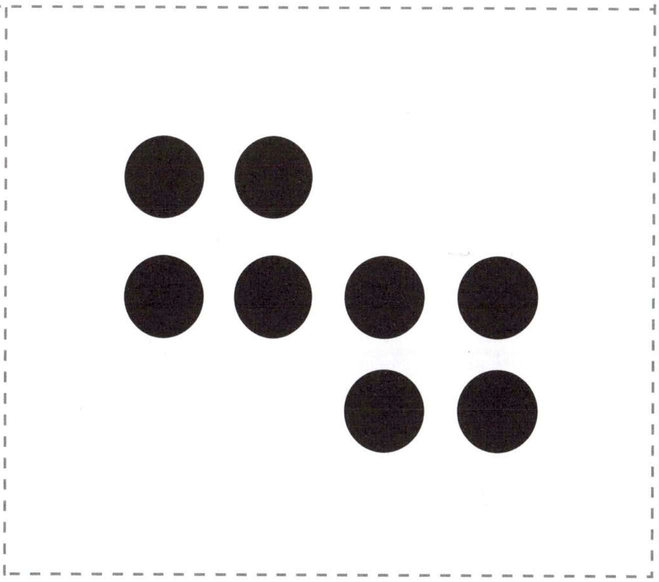

Contemplate Then Calculate

Count and Write How Many | FLUENCY SHEET 1

Count and write how many.

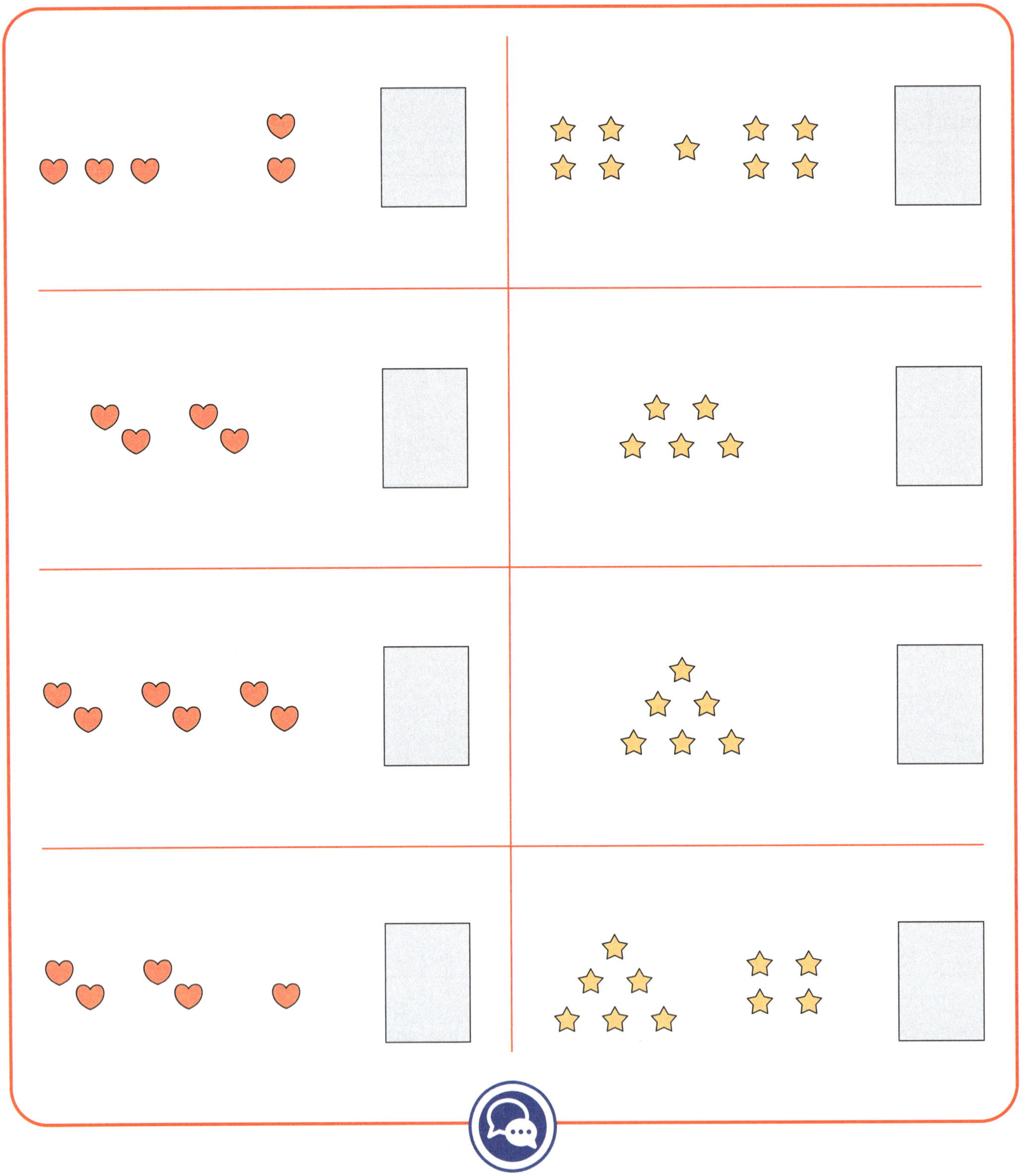

Contemplate Then Calculate

Count and Write How Many | FLUENCY SHEET 2

Count and write how many.

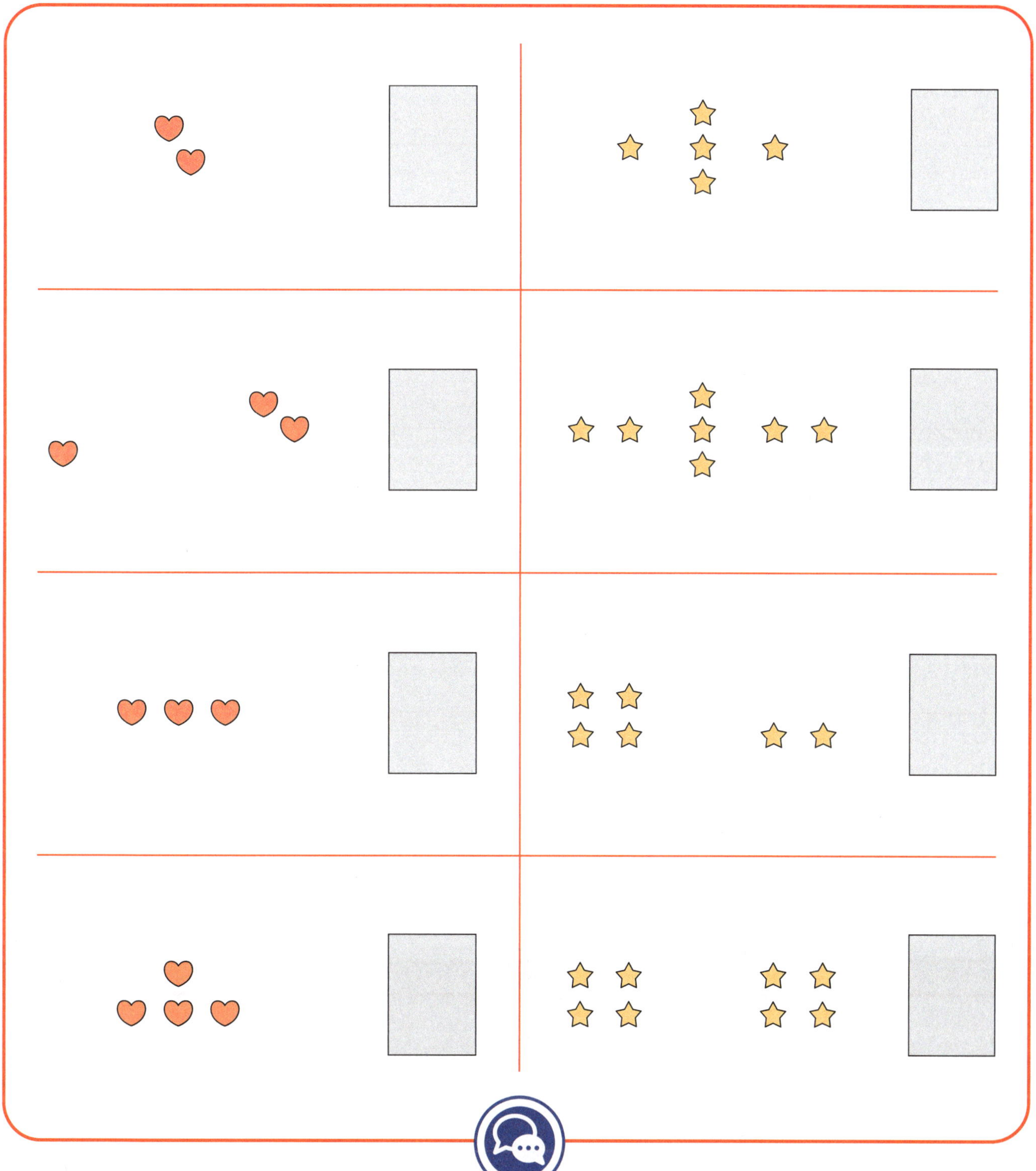

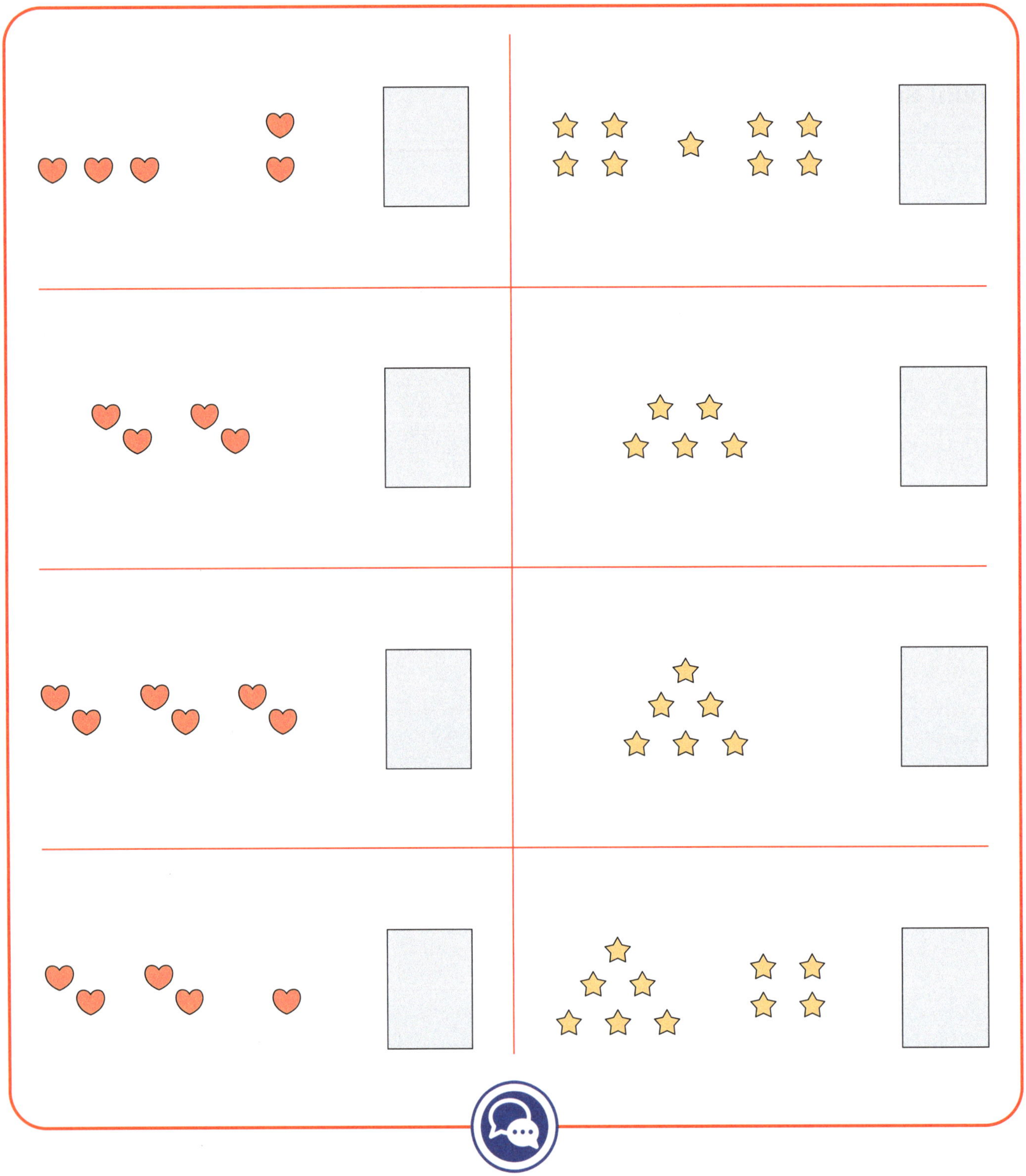

TOPIC A

LESSON 1

Name ____

PROBLEM SET

I can tell or draw an addition story. **I can** write a number sentence to match.

California CCSS-M
K.OA.A.1

Tell a story. Write the number sentence.

3 and 1 make ☐.

3 + 1 = ☐

☐ and ☐ make ☐.

☐ + ☐ = ☐

REMOVABLE

Shake Those Disks

1	2	3	4	5	6	7	8	9	10

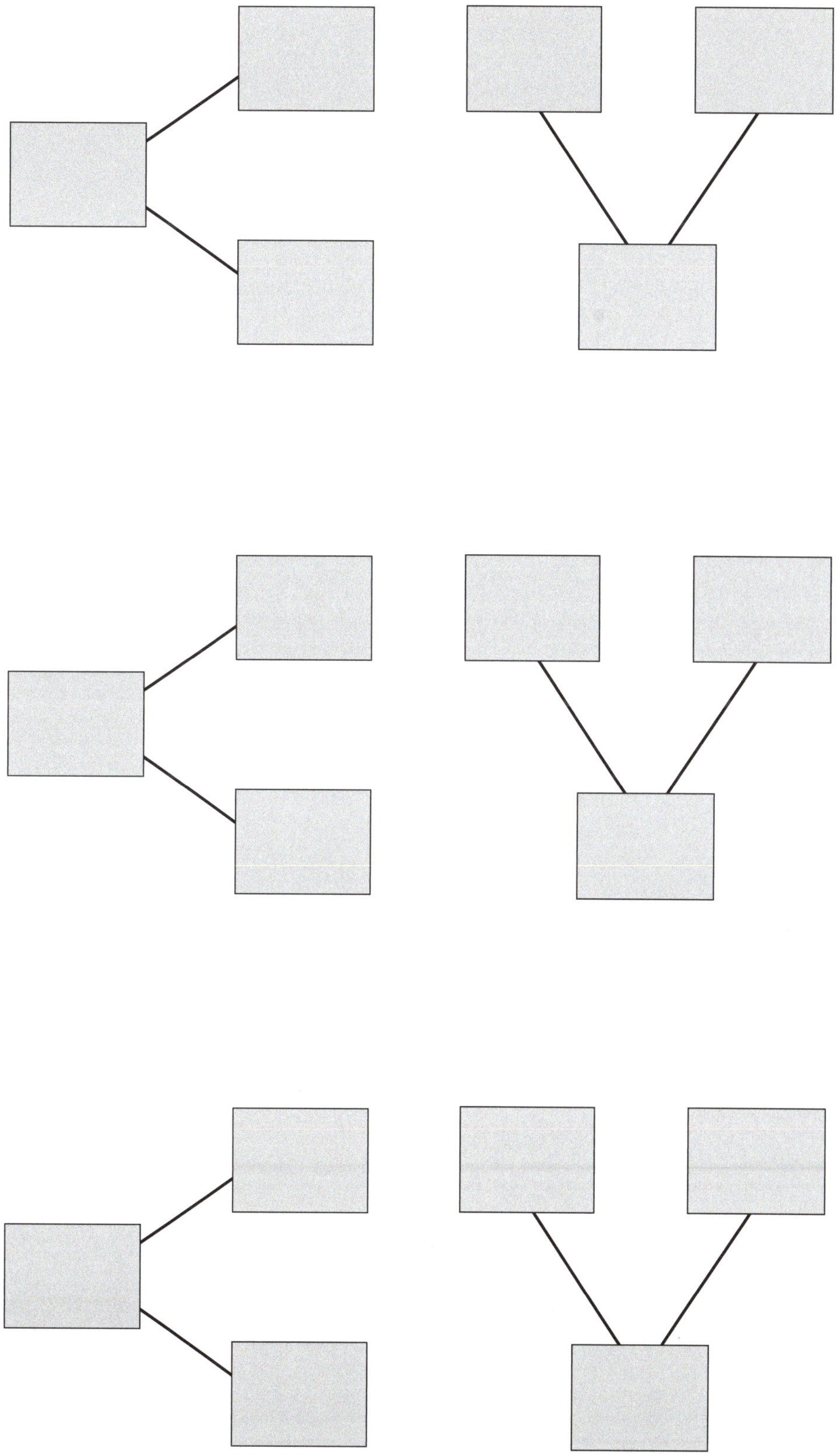

Name ______________________________

PROBLEM SET

I can tell an addition story. **I can** show the parts and total with a number bond and a number sentence.

California CCSS-M
K.OA.A.1

Tell a story. Write the number bond and number sentence.

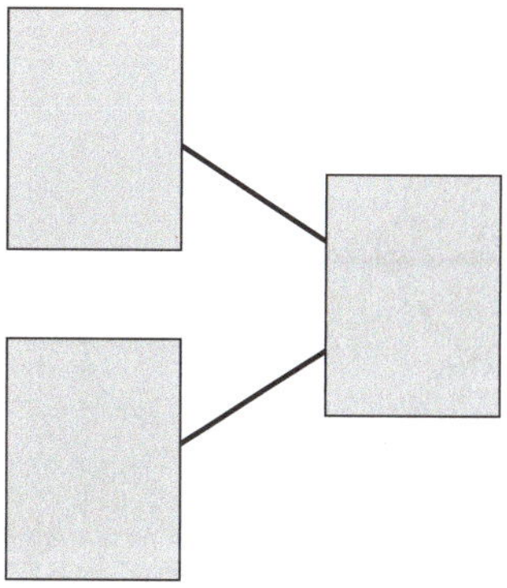

☐ + ☐ = ☐

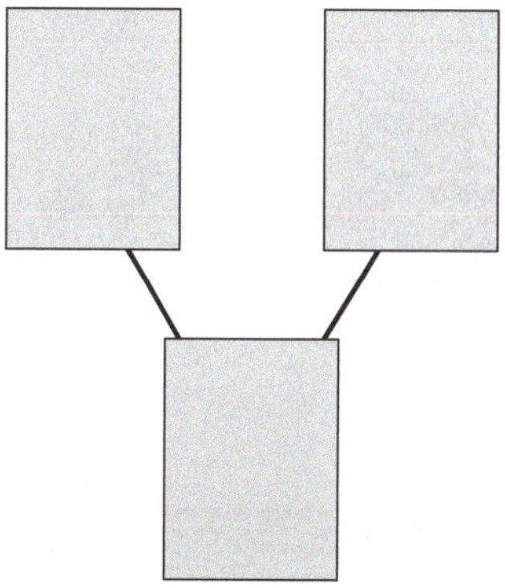

Tell a story. Write the number sentence.

Number Sentence Hunt

REMOVABLE

Walk to a picture. Tell a story. Write the number sentence.

Name ______________________

LESSON

I can tell an addition story. **I can** choose tools to help me solve it. **I can** write a number sentence to match.

California CCSS-M
K.OA.A.1, K.OA.A.2

There are 4 children on the roller coaster. 3 children get on. How many children are on the roller coaster now?

Recording Sheet

REMOVABLE

Sort. Write a number bond and number sentence.

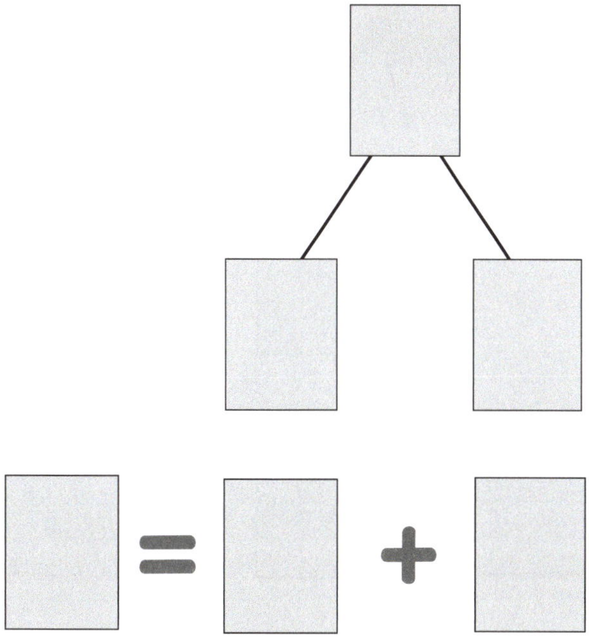

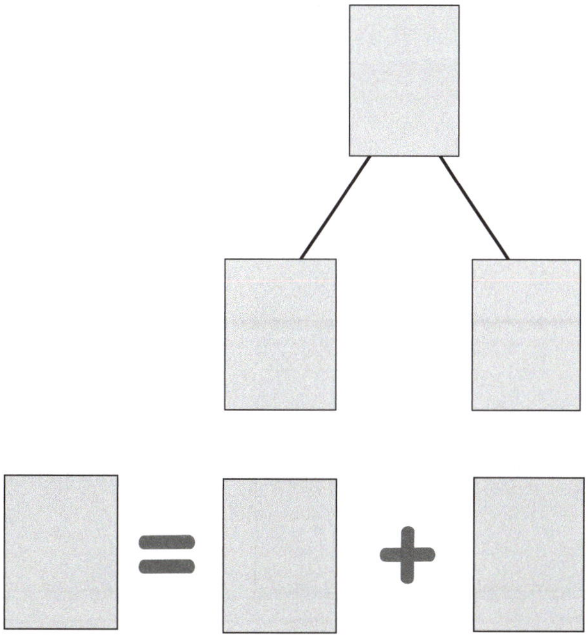

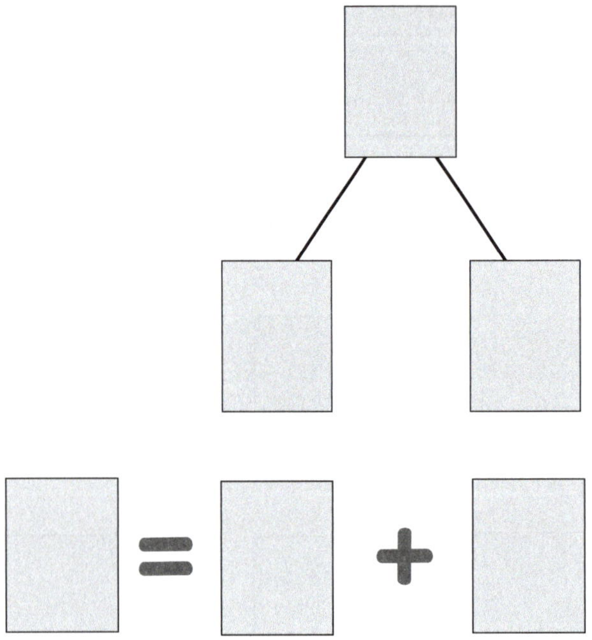

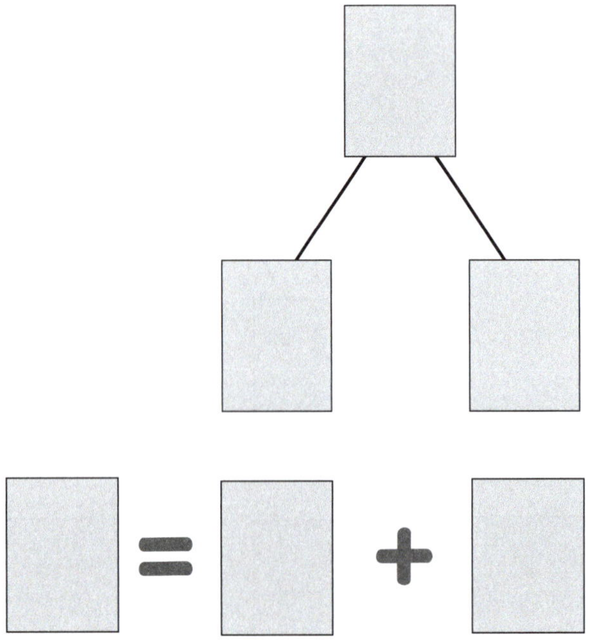

Name ______________________________

PROBLEM SET

I can sort the same total into different parts. **I can** show the parts and total with a number bond and a number sentence.

California CCSS-M
K.OA.A.1, K.OA.A.3

Find the parts. Fill in the number sentence.

5 is ___ and ___.

5 = ___ + ___

7 is ___ and ___.

7 = ___ + ___

$9 = \square + \square$

$\square = \square + \square + \square$

$\square$

Contemplate Then Calculate

Make It Equal | FLUENCY SHEET 1

Draw more dots to make it equal.

3 = 3

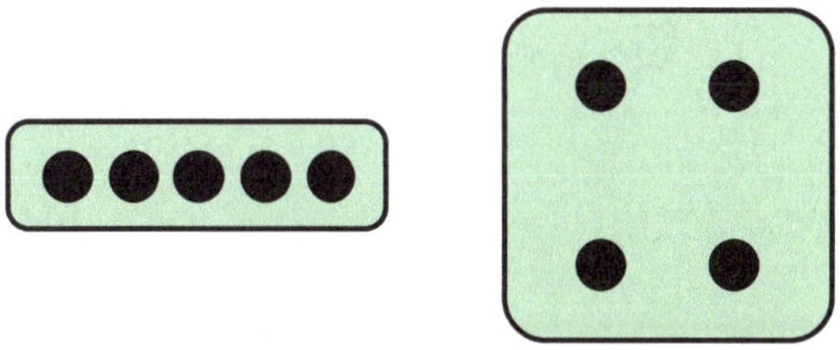

5 = 5

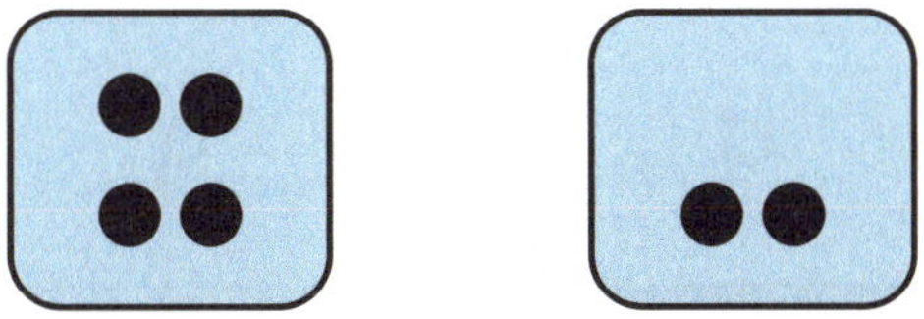

4 = 4

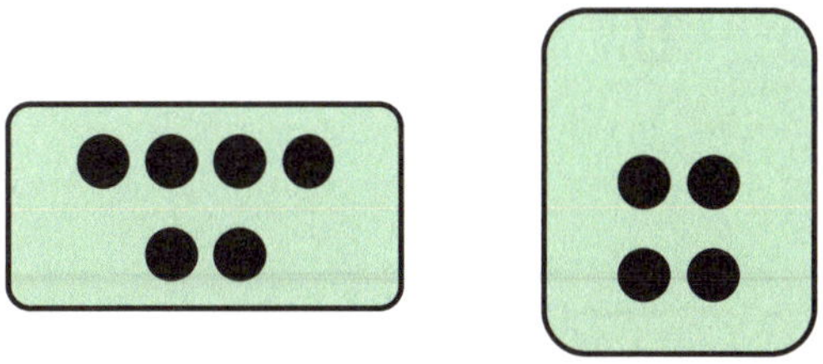

6 = 6

6 = 6

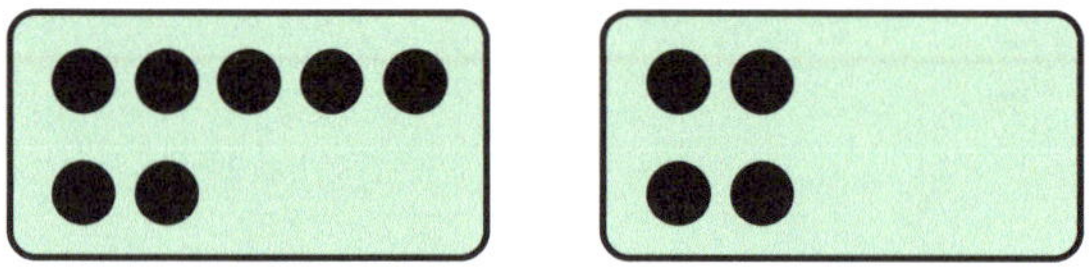

7 = 7

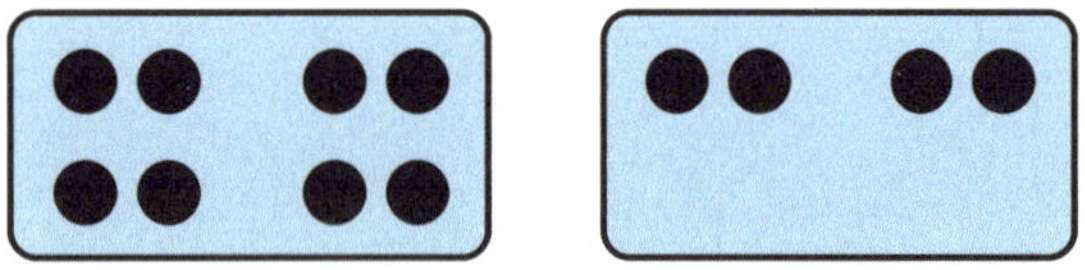

8 = 8

9 = 9

Contemplate Then Calculate

Make It Equal | FLUENCY SHEET 2

Draw more dots to make it equal.

3 = 3

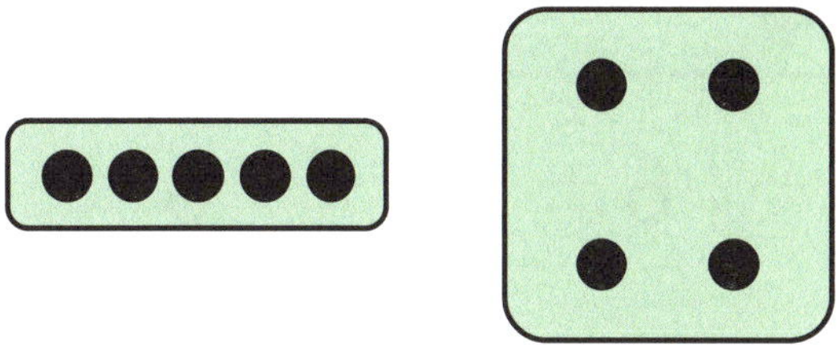

5 = 5

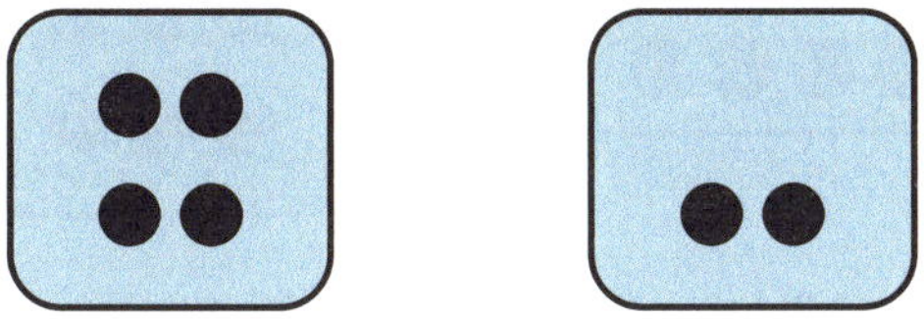

4 = 4

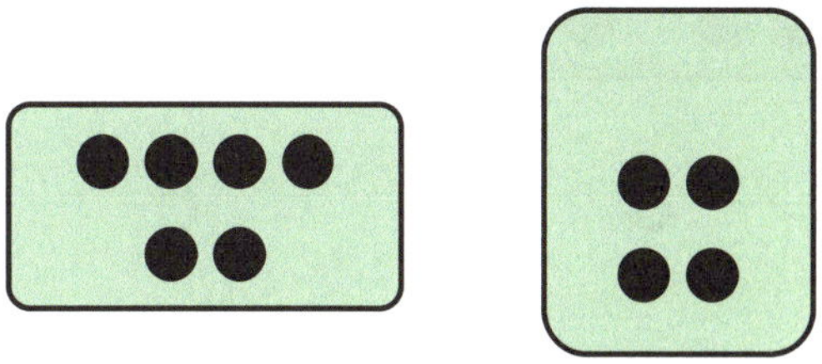

6 = 6

6 = 6

7 = 7

8 = 8

9 = 9

Name ______________________________

LESSON

I can break the total in a math story into parts. **I can** show the parts and total with a number sentence.

California CCSS-M
K.OA.A.1

Listen to the story. Draw. Fill in the number sentence.

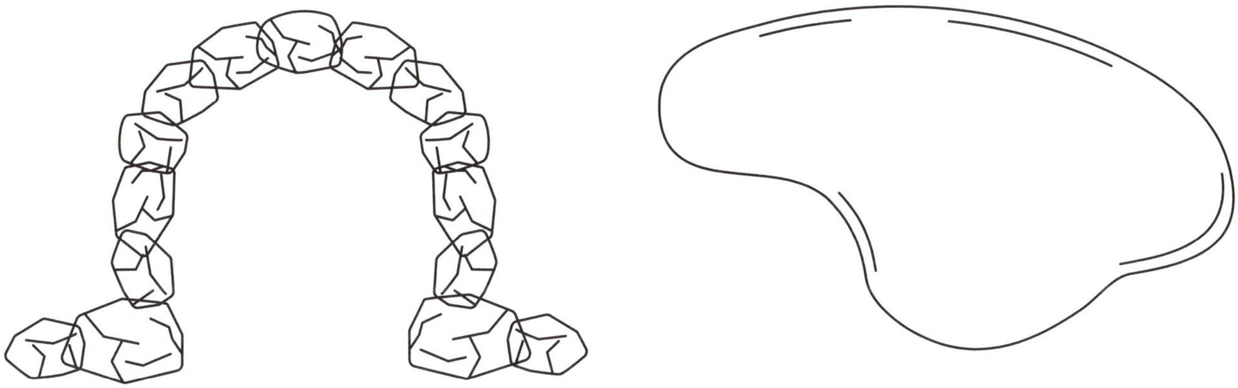

7 = ☐ + ☐

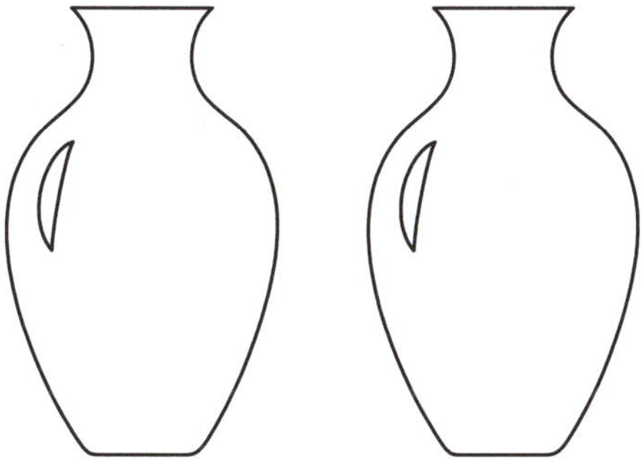

6 = ☐ + ☐

$10 = \square + \square + \square$

TOPIC A

LESSON 5

California CCSS-M
K.OA.A.1

Name ______________________________

PROBLEM SET

I can break the total in a math story into parts. **I can** show the parts and total with a number sentence.

Color to make the parts. Fill in the number sentence.

6 = ☐ + ☐

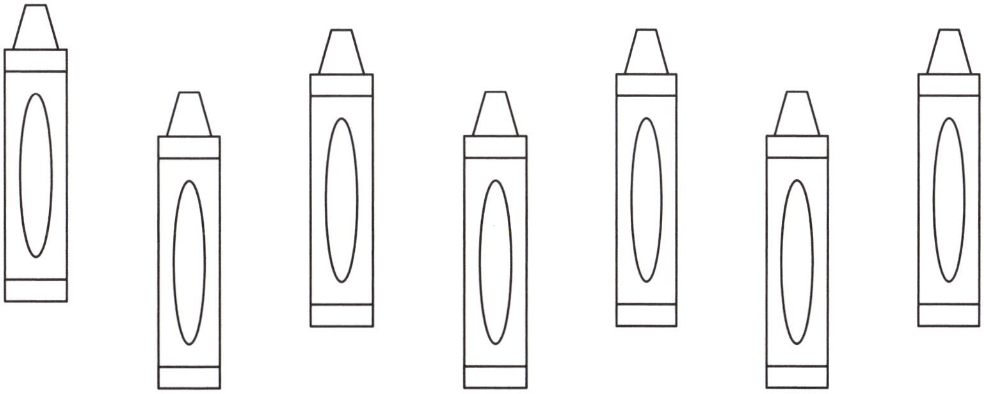

☐ = ☐ + ☐

$7 = \square + \square$

$\square = \square + \square + \square$

Name ______________________

PROBLEM SET

I can choose a tool to help me find a total. **I can** write an addition sentence with that total.

California CCSS-M
K.OA.A.1, K.OA.A.5

Fill in the number sentences.

$1 + 2 = \square$

$2 + 2 = \square$

$3 + 2 = \square$

$4 + 1 = \square$

$5 + 0 = \square$

___ = 5 + 2

___ = 3 + 4

___ = 6 + 3

8 = ___ + ___

___ = 5 + 3 + 2

9 = ___ + ___ + ___

FAMILY MATH

Represent Subtraction

Dear Family,

Students solve subtraction problems and use number sentences to represent their thinking. Actions such as leaving, giving, or eating clearly signal subtraction: "I had 5 watermelon slices. I ate 2 of them. Now there are 3 watermelon slices." Students use the familiar story structure of beginning, middle, and end to write and understand number sentences.

Key Terms

Subtraction

Minus

Students transition from reading number sentences as "9 **take away** 5 **is** 4" to "9 **minus** 5 **equals** 4."

9 take away 5 is 4.

$9 - 5 = 4$

At-Home Activities

Activity Idea 1 Missing Pennies

Gather 10 pennies and a small container. Place the pennies on a table or workspace. Ask your child to hide some of the pennies in the container while you look away. When they're ready, work together to write a number sentence to match what happened. Consider asking the following questions:

- How many pennies did we start with?
- What happened next? How could we write that?
- How many pennies are left?

Play again with different number combinations and take turns hiding the pennies.

Activity Idea 2 Bowling at Home

Set up a bowling game at home. Invite your child to set up 10 household items as pins. The items could be toys, spice containers, markers, or plastic bottles. Ask your child to count the items to make sure there are 10. Invite them to roll a ball to knock down the items.

Ask, “How many pins fell?” and “How many are still standing?” Your child can write a subtraction sentence to match what happened. For example, “We started with 10 bottles. I knocked down 3. There are 7 left. I can write that as 10 – 3 = 7.”

Subtraction Number Sentences

In addition problems, each part plays the same role. They are combined to make the total. As students move to subtraction, they realize that each part plays a different role. One is the part that is removed and the other is the part that remains. Phrases such as *take away* and *are left* are useful to help distinguish between the parts.

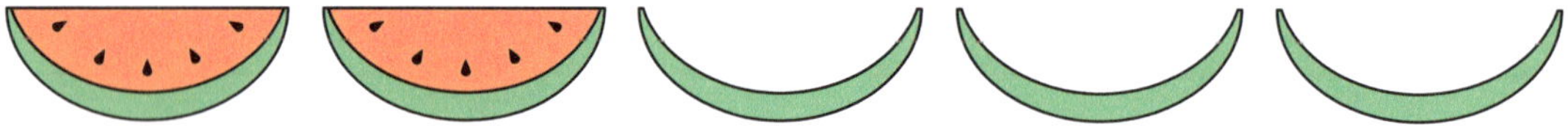

5 take away 3 is ☐.

$5 - 3 = \square$

Students can also identify the parts and total in a subtraction problem by sorting. They begin with the total and sort to take one part away.

There are 7 crayons at the writing center. Someone takes 4 blue crayons.
How many crayons are left?

$7 - 4 = 3$

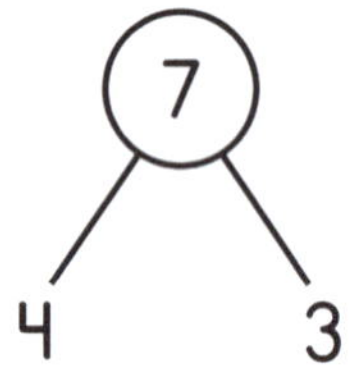

Special attention is placed on identifying the part that went away and the part that is left in the number sentence. Students see that number bonds do not show this information, but number sentences do.

Apple Tree

REMOVABLE

☐ apples

Name ______________________________

PROBLEM SET

I can take some away and write how many are left.

California CCSS-M
K.OA.A.1

Mouse ate 2 of each snack. Write how many are left.

Mouse ate 2 of each snack. Write how many are left.

Name ______________________________

PROBLEM SET

I can tell or draw a subtraction story.
I can write a number sentence to match.

California CCSS-M
K.OA.A.1

Tell a story using the picture. Fill in the number sentence.

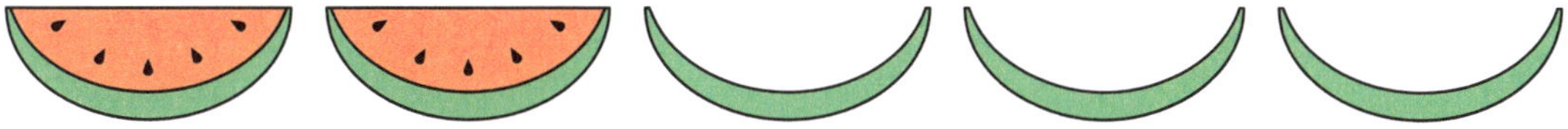

5 take away 3 is ☐.

5 − 3 = ☐

8 take away 2 is ☐.

8 − ☐ = ☐

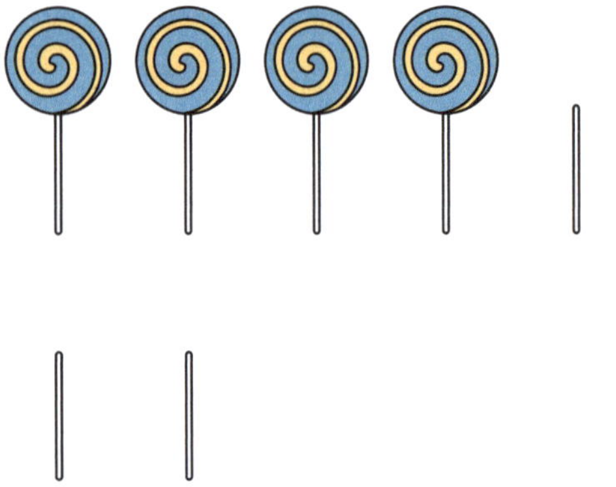

7 − ☐ = ☐

Contemplate Then Calculate

Take Away 1 | FLUENCY SHEET 1

Cross out 1 and write how many are left.

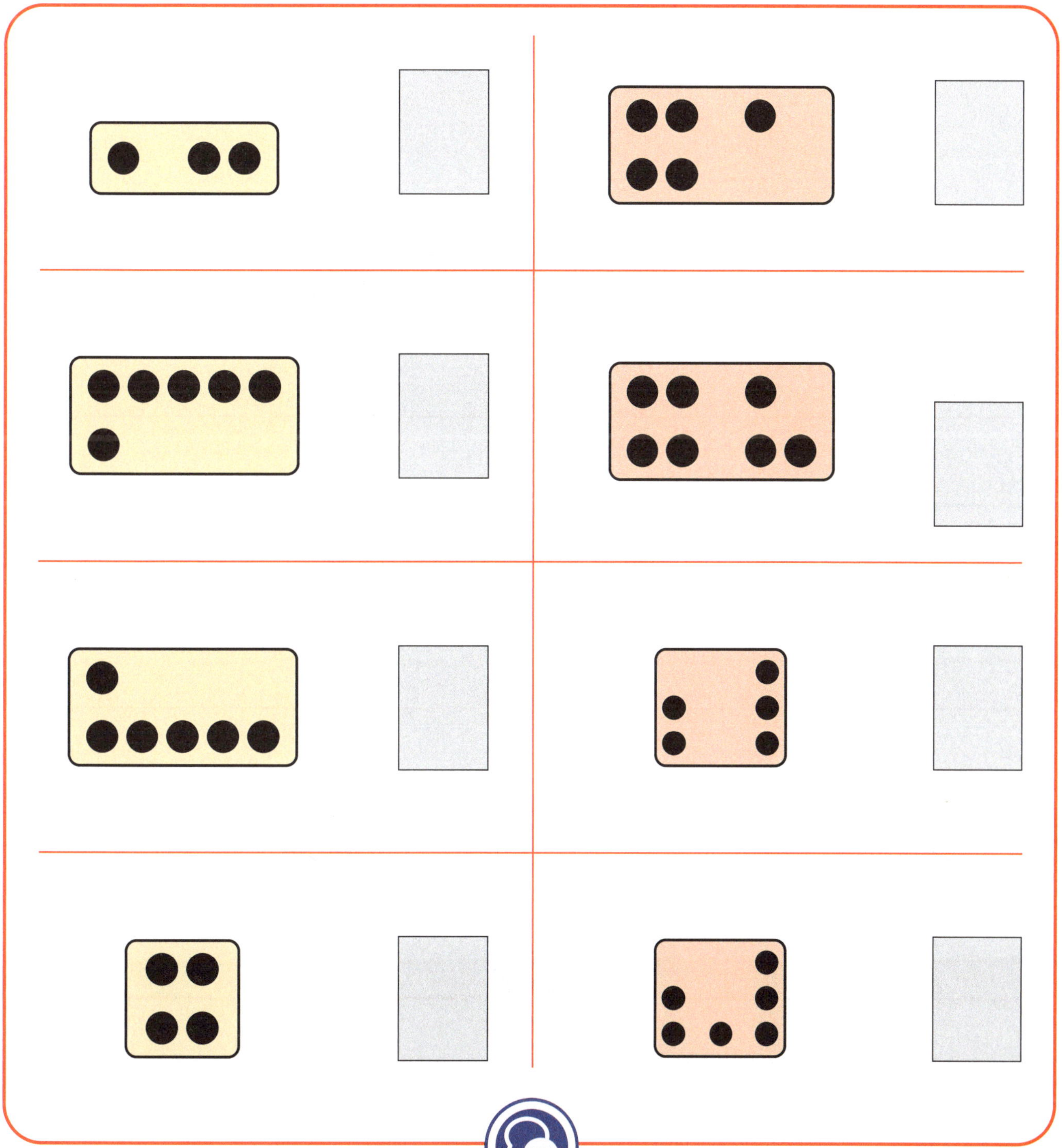

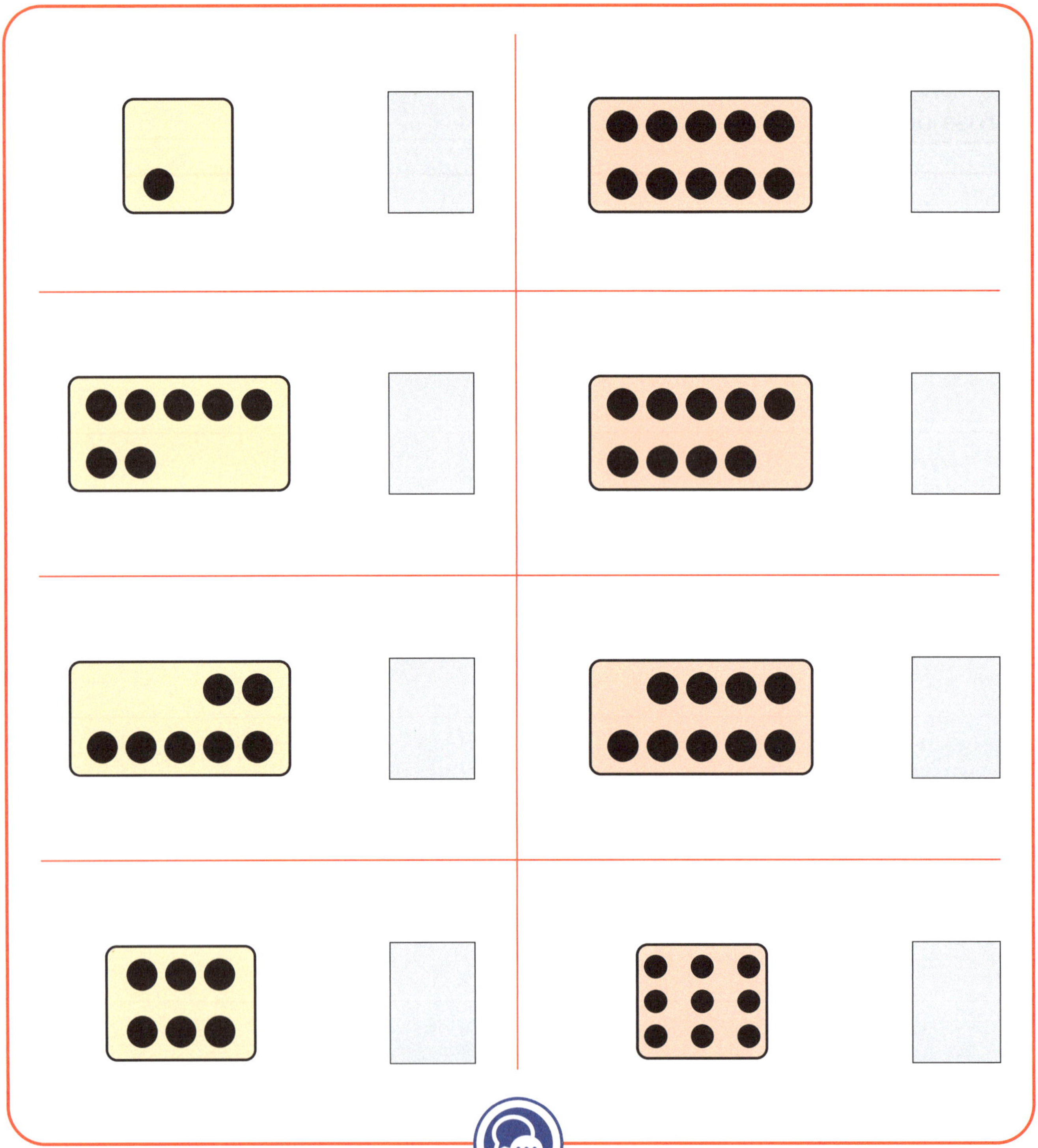

Contemplate Then Calculate

Take Away 1 | FLUENCY SHEET 2

Cross out 1 and write how many are left.

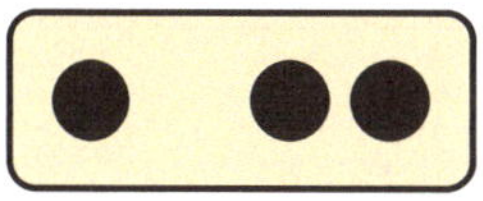

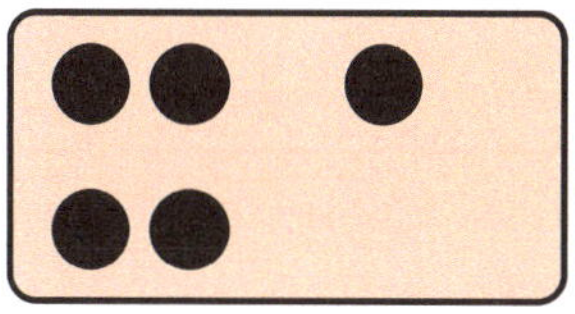

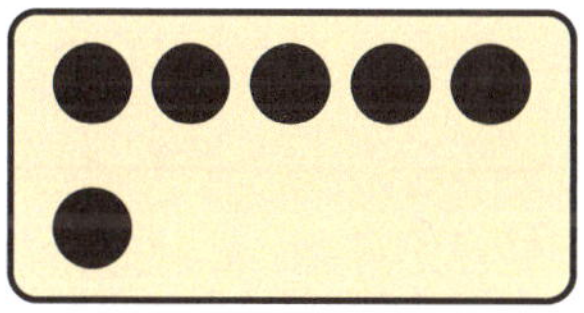

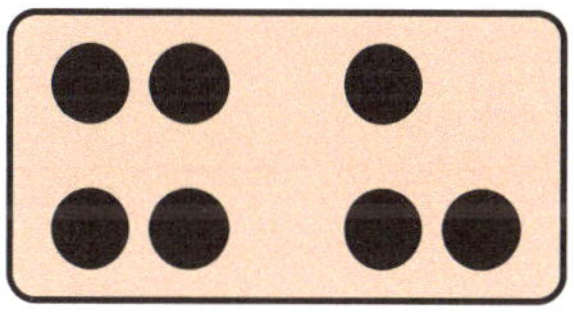

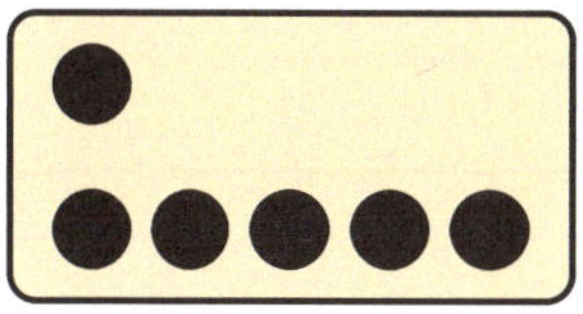

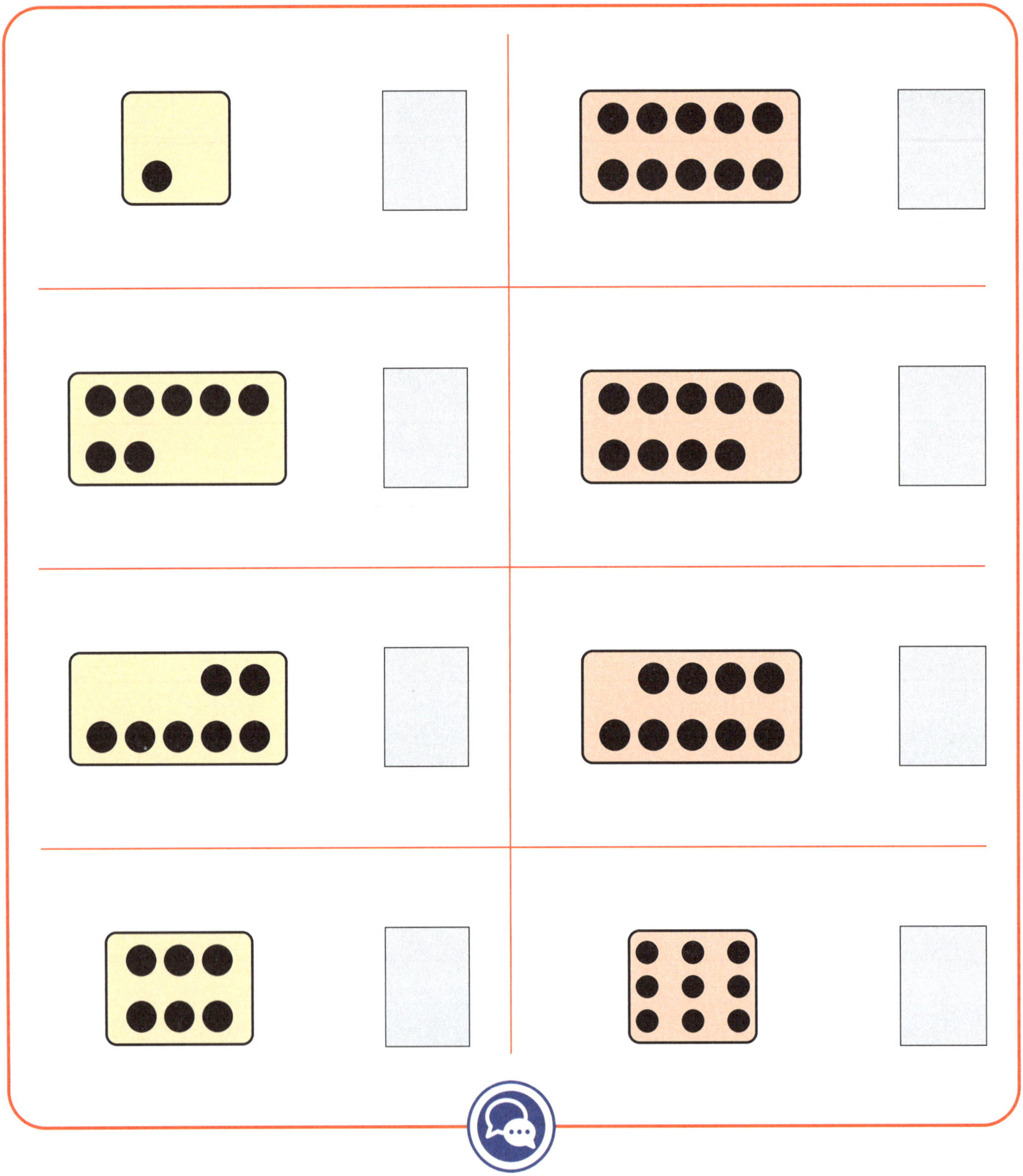

Name ______________________________

LESSON

I can choose a tool to help me find how many are left. **I can** write a number sentence to match.

California CCSS-M
K.OA.A.1, K.OA.A.2

Edwin has 7 cookies. He eats 2 cookies. How many cookies does Edwin have now?

Name ______________________________

PROBLEM SET

I can choose a tool to help me find how many are left. **I can** write a number sentence to match.

California CCSS-M
K.OA.A.1, K.OA.A.2

Mr. Triangle takes away all the triangles. Fill in the number sentence.

$8 - 2 = \square$

$9 - \square = \square$

☐ − ☐ = ☐

Draw some shapes. Cross out the triangles. Fill in the number sentence.

☐ − ☐ = ☐

Number Bond

REMOVABLE

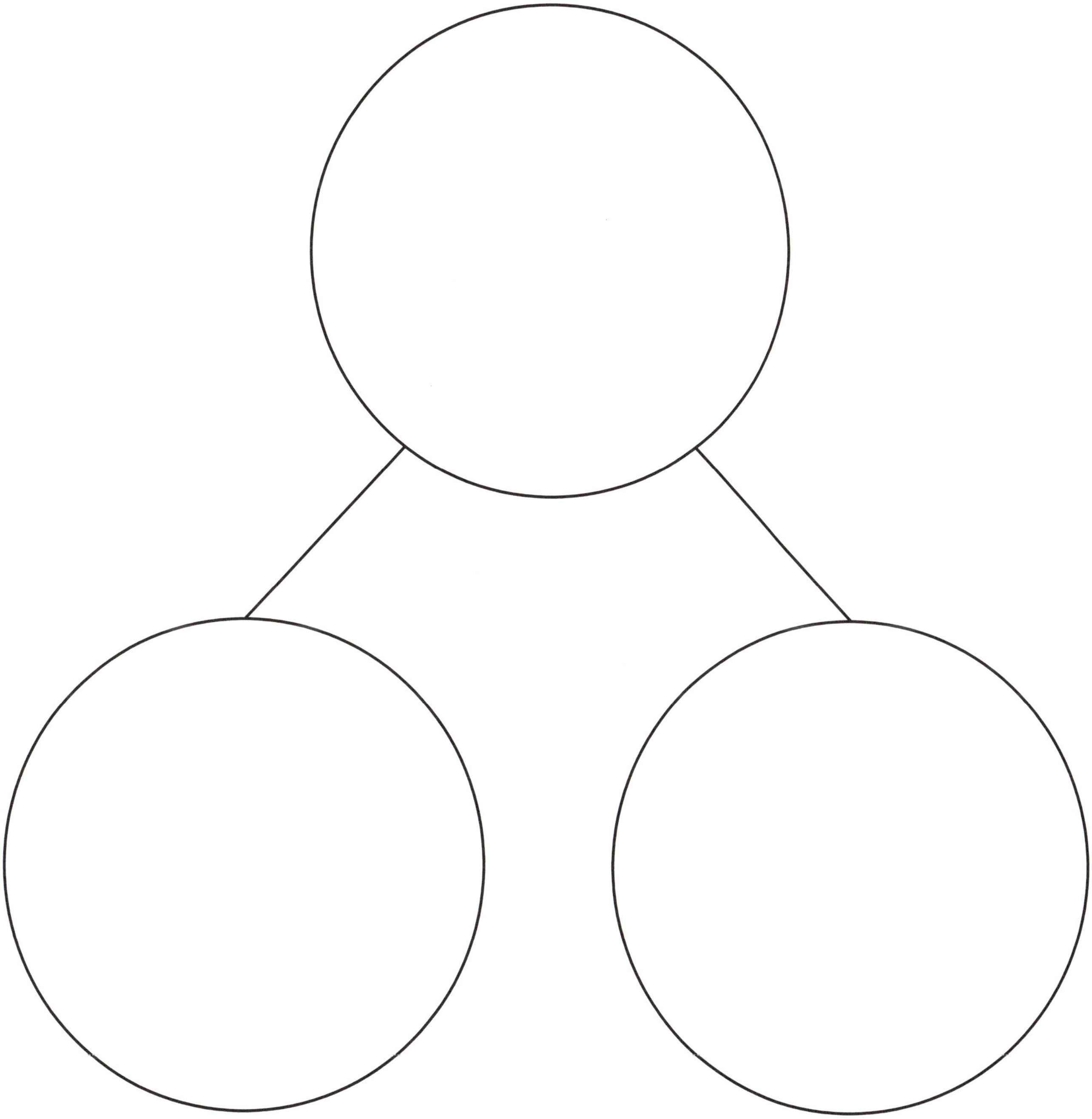

1	2	3	4	5	6	7	8	9	10

Number Bond

REMOVABLE

1	2	3	4	5	6	7	8	9	10

Name ______________________________

LESSON

I can sort the same total into different parts. **I can** show the parts and total with a number bond and a subtraction sentence.

California CCSS-M
K.OA.A.1

Recording Sheet

Sort. Write a number bond and number sentence.

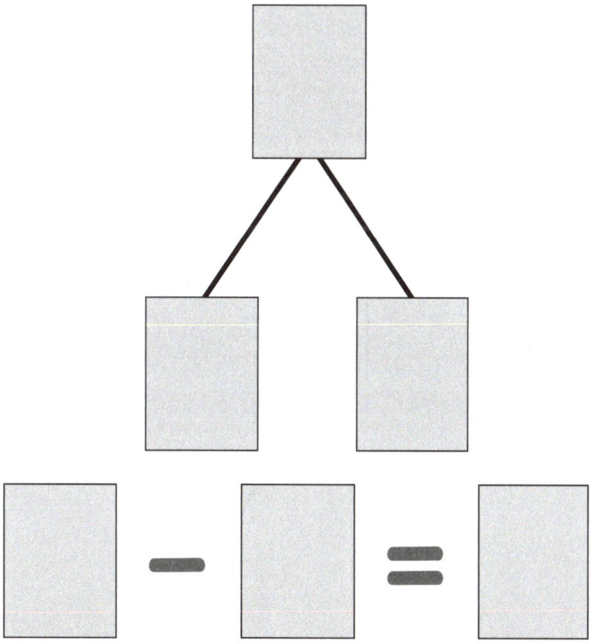

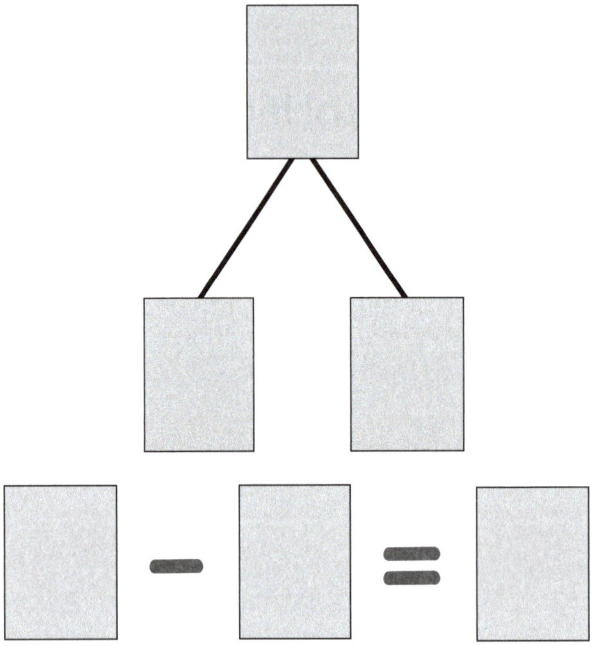

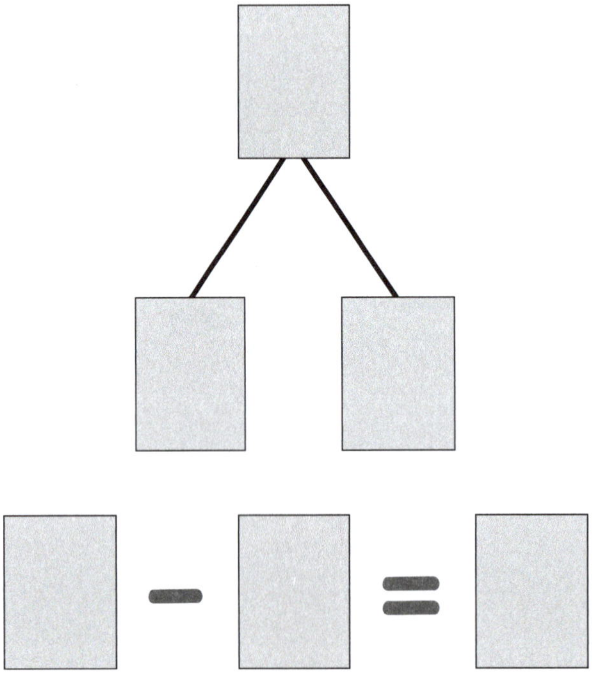

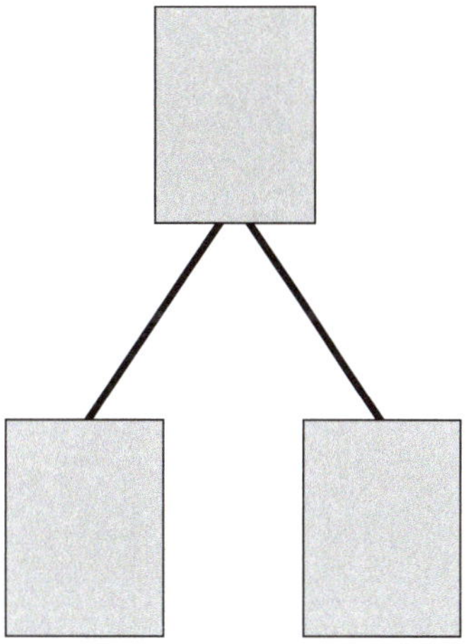

$\square - \square = \square$

Name ______________________________

California CCSS-M
K.OA.A.1

PROBLEM SET

I can sort the same total into different parts. **I can** show the parts and total with a number bond and a subtraction sentence.

Cross out a part. Fill in the number sentence.

8 − ☐ = ☐

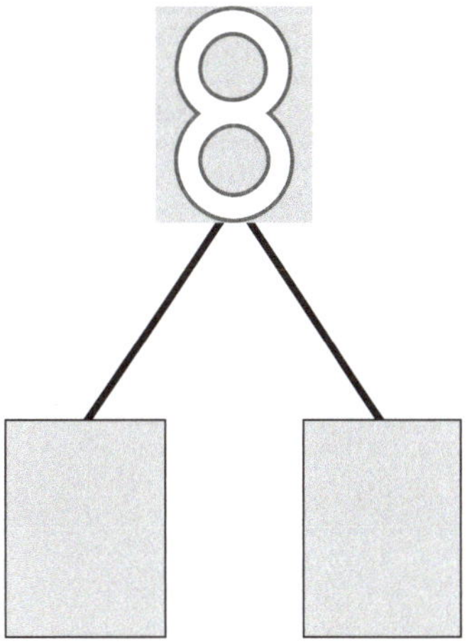

Cross out a part. Fill in the number sentence.

8 − ☐ = ☐

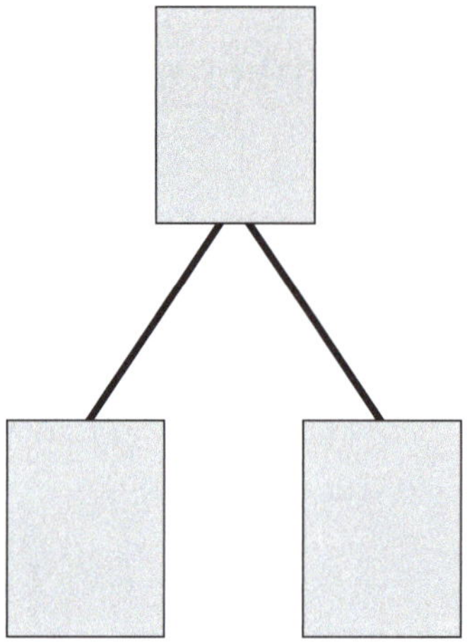

Cross out a part. Fill in the number sentence.

☐ − ☐ = ☐

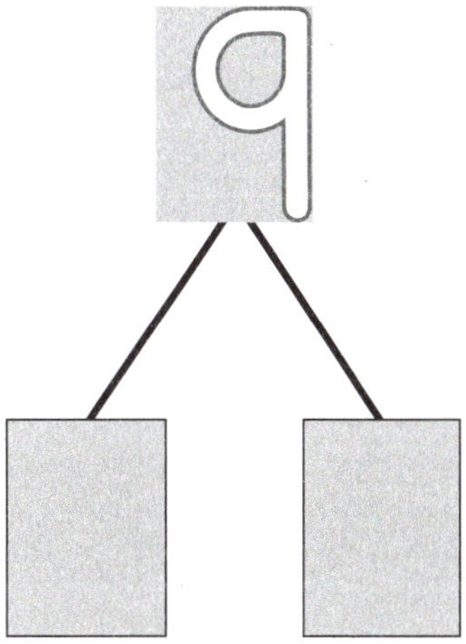

☐ − ☐ = ☐

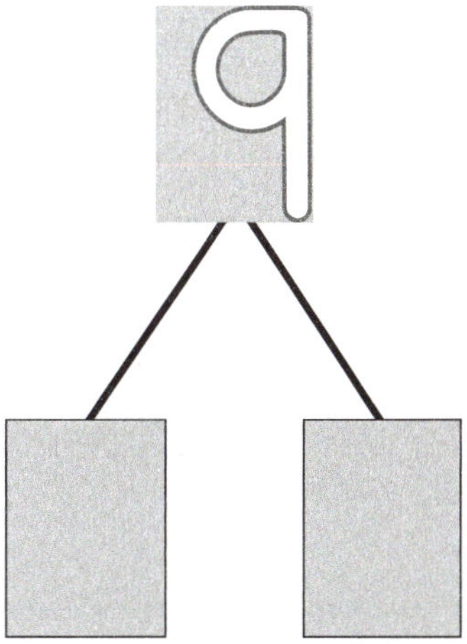

Bowling Recording Sheet

REMOVABLE

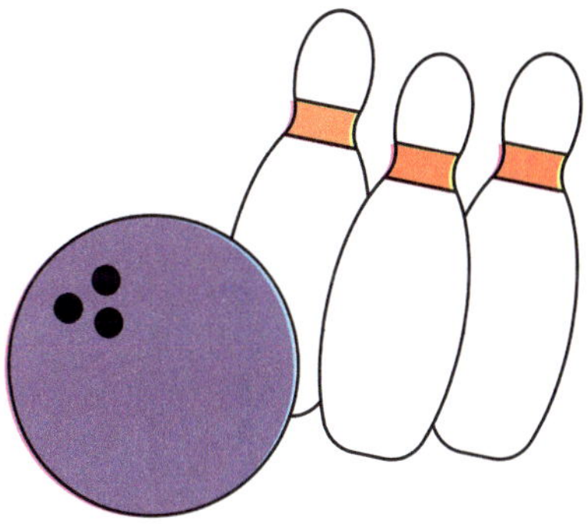

10 − ☐ = ☐

10 − ☐ = ☐

☐ − ☐ = ☐

☐ − ☐ = ☐

Name ______________________________

LESSON

I can choose a tool to help me find how many are left. **I can** show the parts and total with a number bond and a subtraction sentence.

California CCSS-M
K.OA.A.1, K.OA.A.2

A chef had 8 eggs. She used 6 of them for breakfast. How many are left?

Subtraction Match

REMOVABLE

Cut. Match the picture to the number sentence.

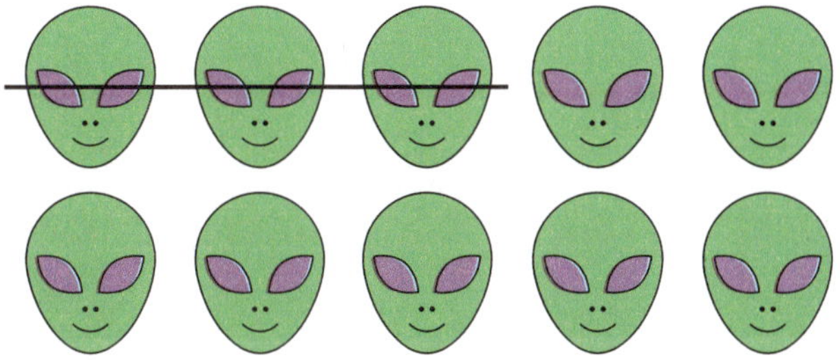
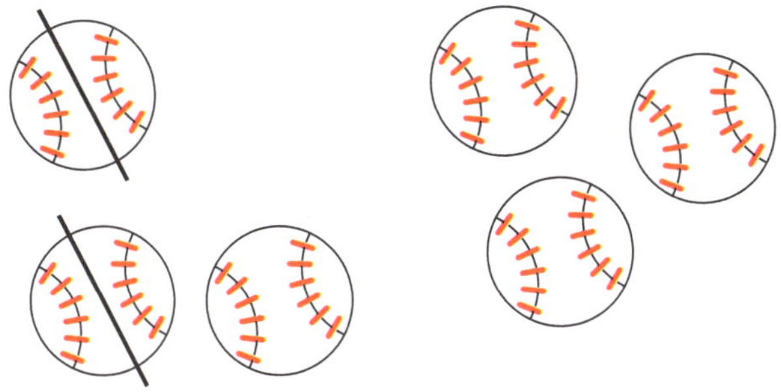

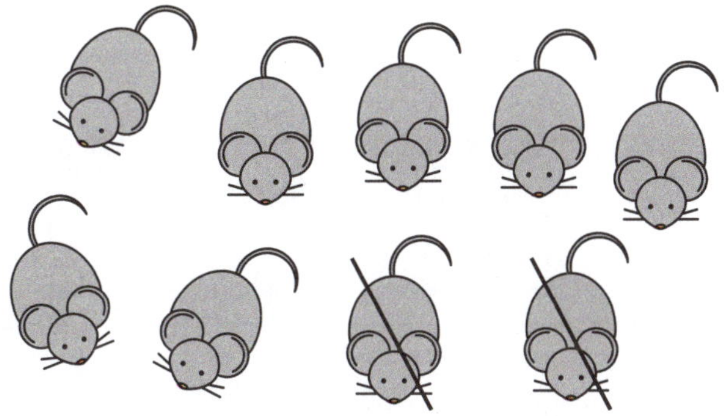

10 - 3 = 7	10 - 0 = 10
8 - 3 = 5	5 - 5 = 0
6 - 2 = 4	4 - 2 = 2
9 - 2 = 7	7 - 5 = 2

Name ______________________________

PROBLEM SET

I can choose a tool to help me find how many are left. **I can** write how many are left in a subtraction sentence.

California CCSS-M
K.OA.A.1, K.OA.A.5

Fill in the number sentence.

3 − 1 = ☐

5 − 4 = ☐

6 − 3 = ☐

4 − 1 = ☐

5 − 0 = ☐

$5 - 1 = \square$

$7 - 4 = \square$

$6 - 6 = \square$

$10 - 4 = \square$

$8 - \square = 2$

$\square - 3 = 2$

FAMILY MATH

Make Sense of Problems

Dear Family,

Students develop the ability to make sense of addition and subtraction by focusing on questions such as, *What is happening in the story?* and *How can I solve this problem?* instead of focusing on keywords (e.g., plus, altogether), which can lead to errors as problems increase in complexity. Students make sense of problems by acting out or modeling actions with tools. Modeling helps students determine whether to write an addition or subtraction sentence.

Students are given a choice of tools such as cubes, hands, number paths, drawings, or 10-frames to solve problems.

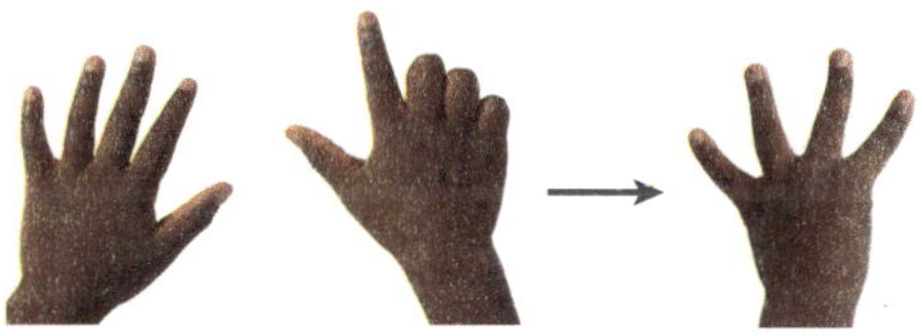

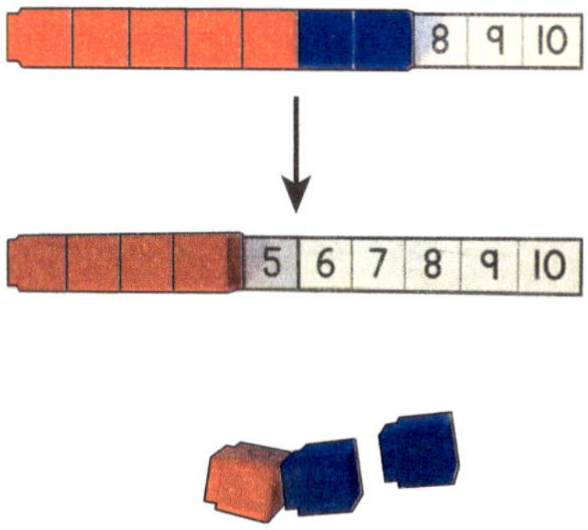

Students can change course at any time by opting to use a different tool. As students learn to solve story problems, they begin to recognize which tools they personally prefer, and which tools work best for different problems.

At-Home Activities

Activity Idea 1 Plus or Minus?

Invite your student to play a game called Plus or Minus. Students listen to a story and decide if they would write an addition or subtraction sentence to match the story. Prompt your child to say “plus” if they would write an addition sentence to match the story or “minus” if they would write a subtraction sentence to match it. Try a story such as the following: 3 birds are eating worms. 5 more birds come to eat worms. Is this plus or minus?

Activity Idea 2 Number Path Game

For this partner game, gather a die and two small household items, such as a penny and a button. Using the number path provided, the goal is to get both items to the number 8. Partner A rolls the die and places their item on the number they roll. Partner B rolls the die and places their item on the number they roll. Partners take turns rolling the die and moving their item the number of spaces that matches the roll. They may move their game piece forward or backward. Play continues until one partner lands on 8. Once you have finished the round, play again with a new target number. If a player is on 5 and rolls a 6, they should roll again.

Making Sense of Problems

Students are beginning to solidify their understanding of when and why to use addition or subtraction. They focus on the action of joining or separating within the story to determine the operation.

When solving a story problem, students

- draw or use tools to represent the story,
- write a number sentence to match the story, and
- explain why they used a plus sign or a minus sign.

In the sample below, students could tell an addition or subtraction story based on the image.

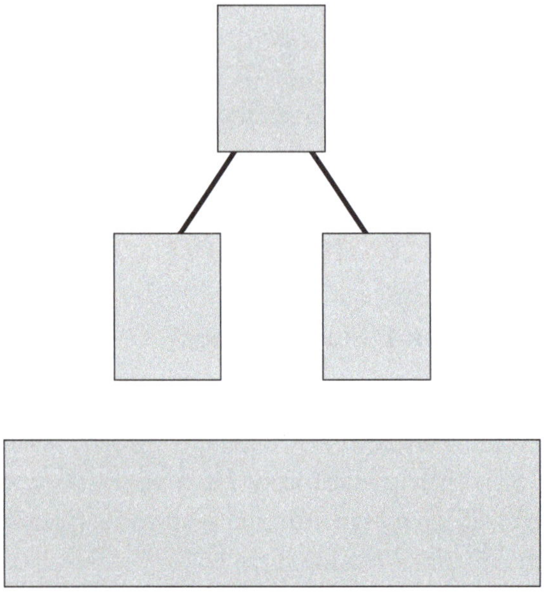

The operation is not named in the story problems (e.g., “addition word problem”) to allow for flexible solving. Students will discover the relationship between addition and subtraction and how one can be used to solve the other. It also allows students to see the reversibility of addition and subtraction, such as in the following story: “There were 5 bunnies and 2 more came along. Now there are 7 bunnies. Later, 2 of them ran away. Now there are 5 once again.”

$$5 + 2 = 7 \text{ and } 7 - 2 = 5$$

Both problems have the same parts and total. Students see that one operation undoes the other.

1	2	3	4	5	6	7	8	9	10

Contemplate Then Calculate

Take Away 2 | FLUENCY SHEET 1

Cross out 2 and write how many are left.

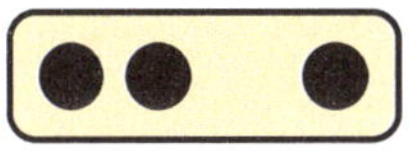

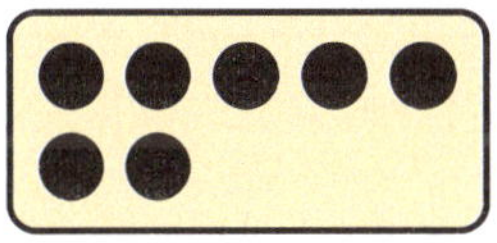

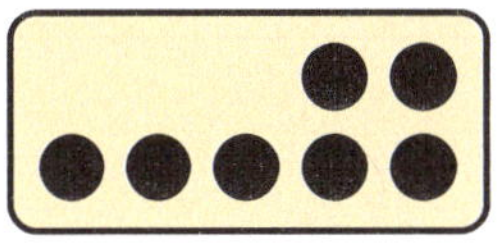

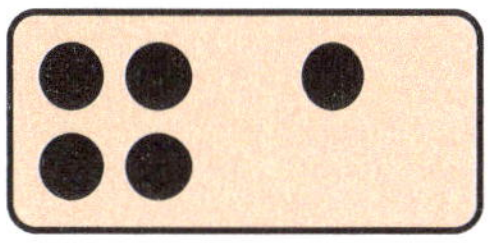

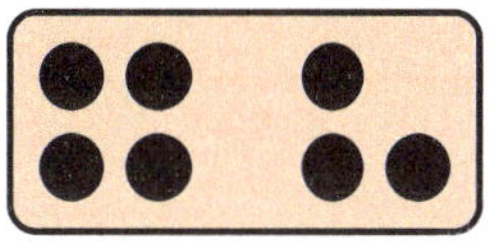

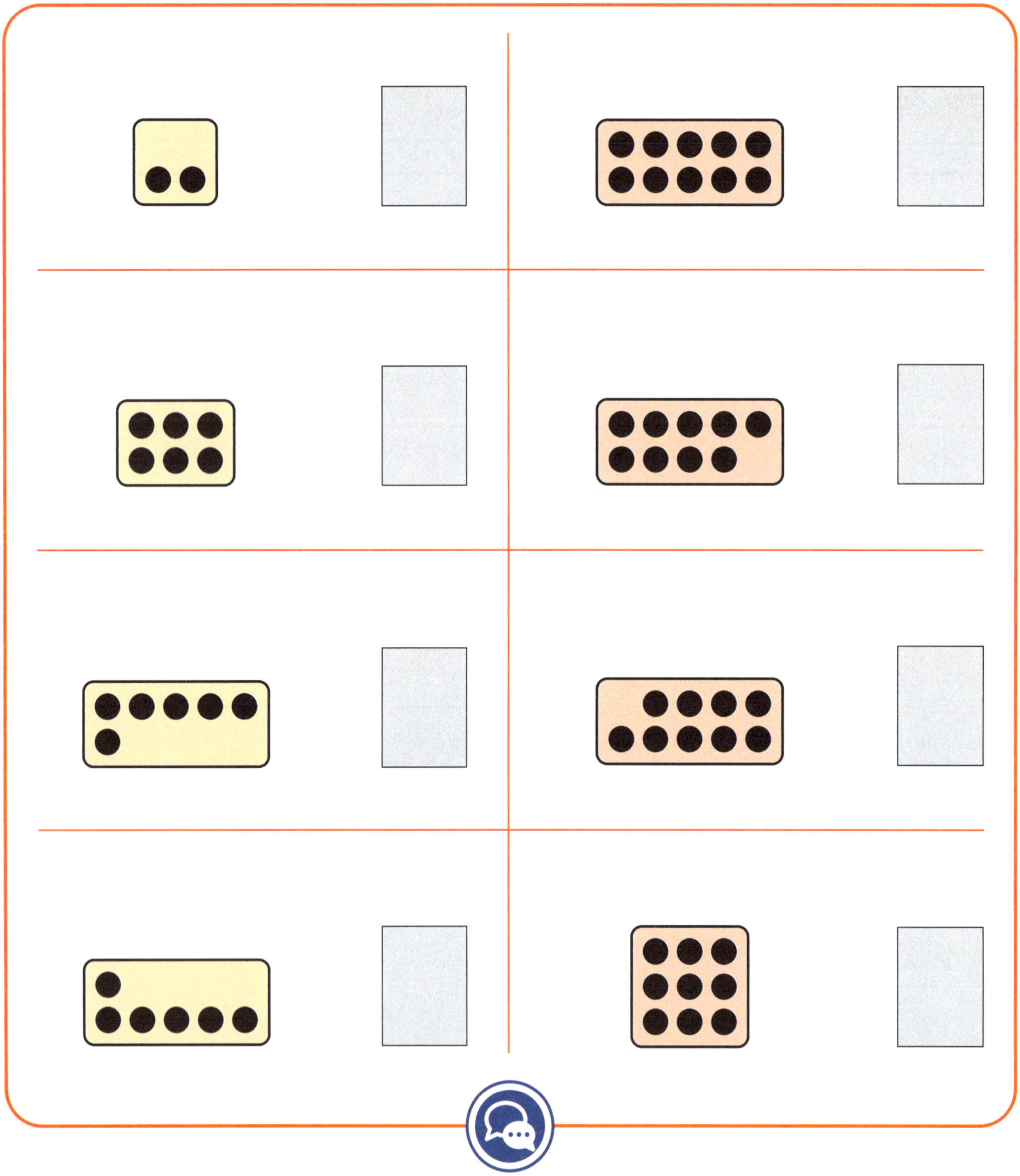

Contemplate Then Calculate

Take Away 2 | FLUENCY SHEET 2

Cross out 2 and write how many are left.

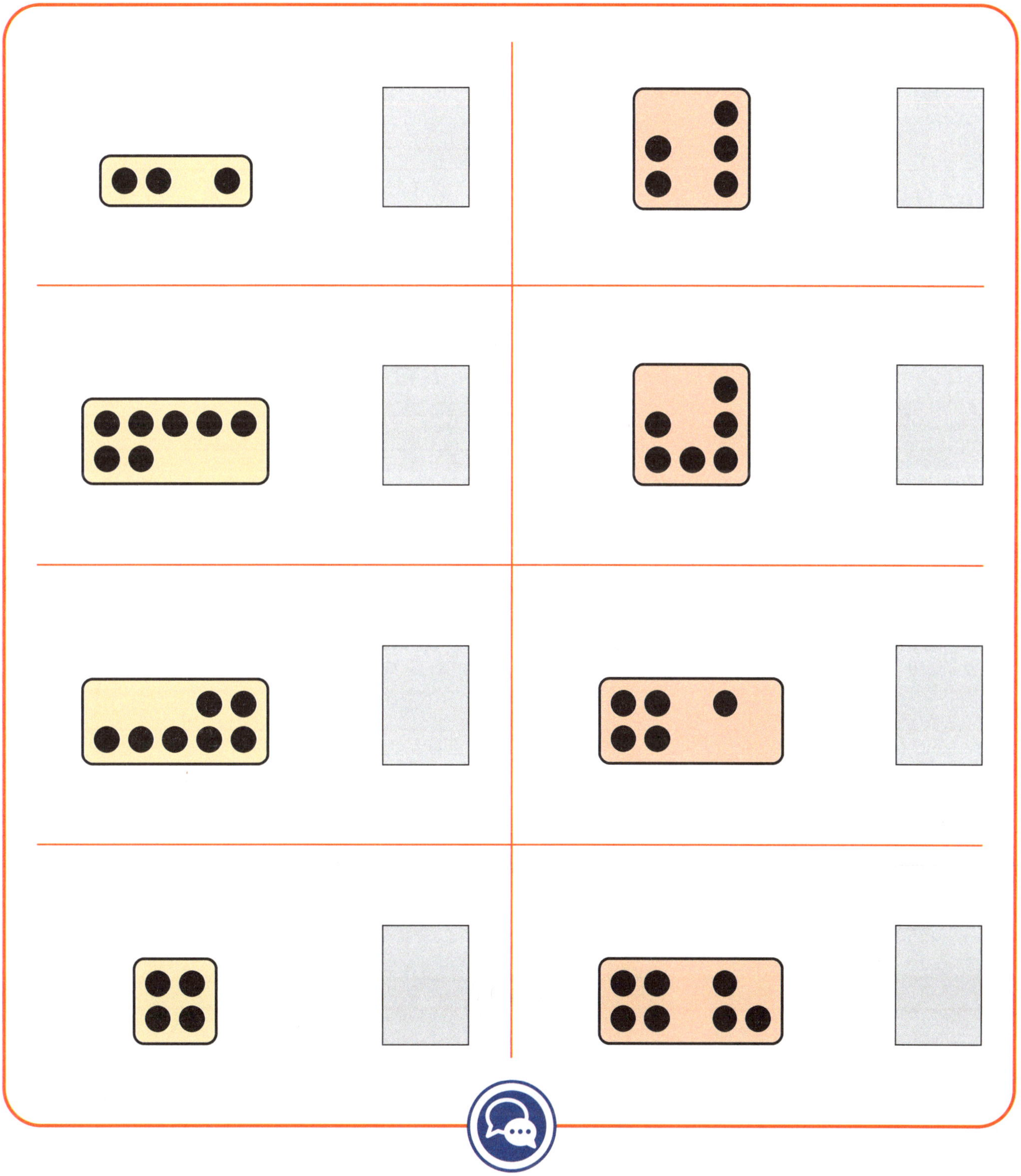

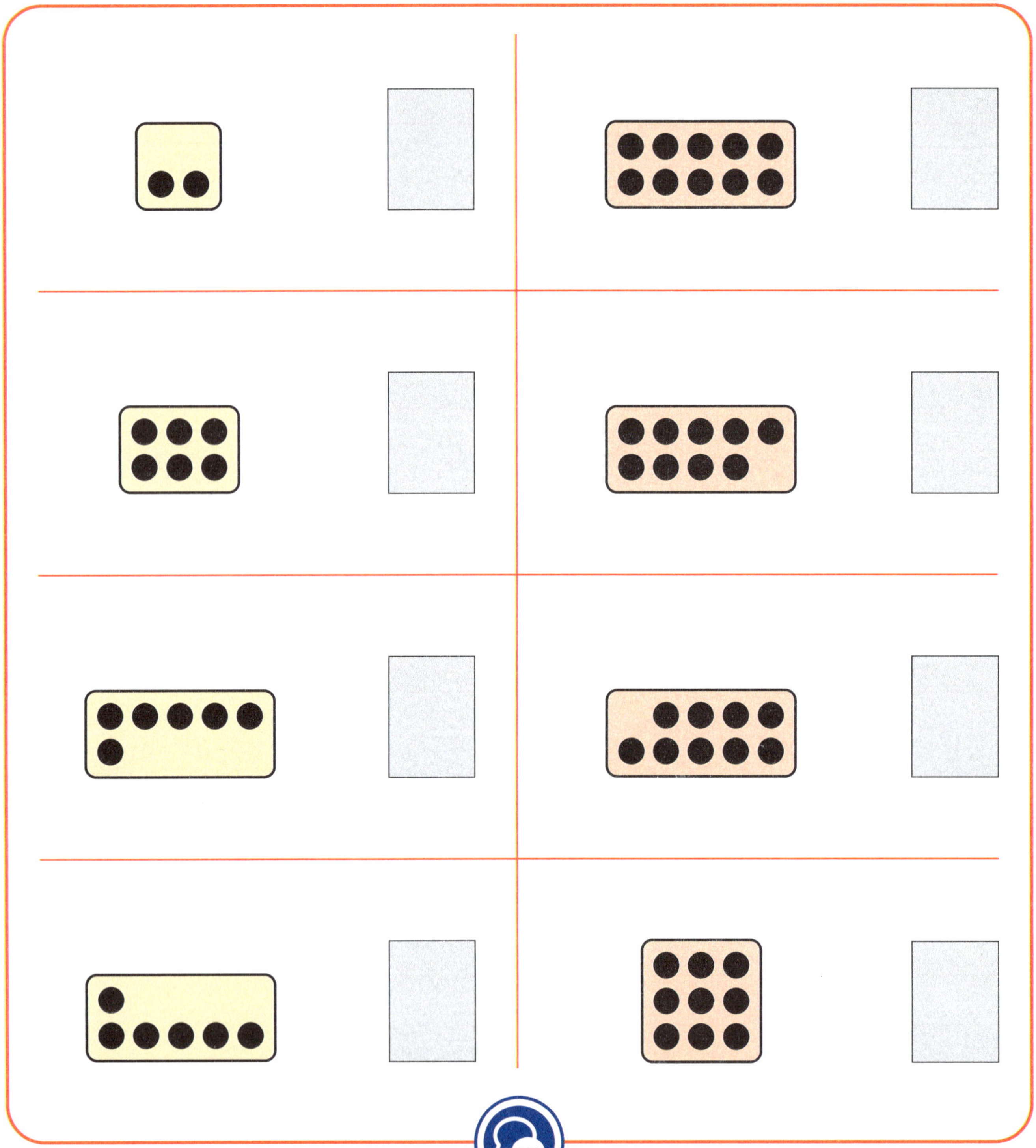

REMOVABLE

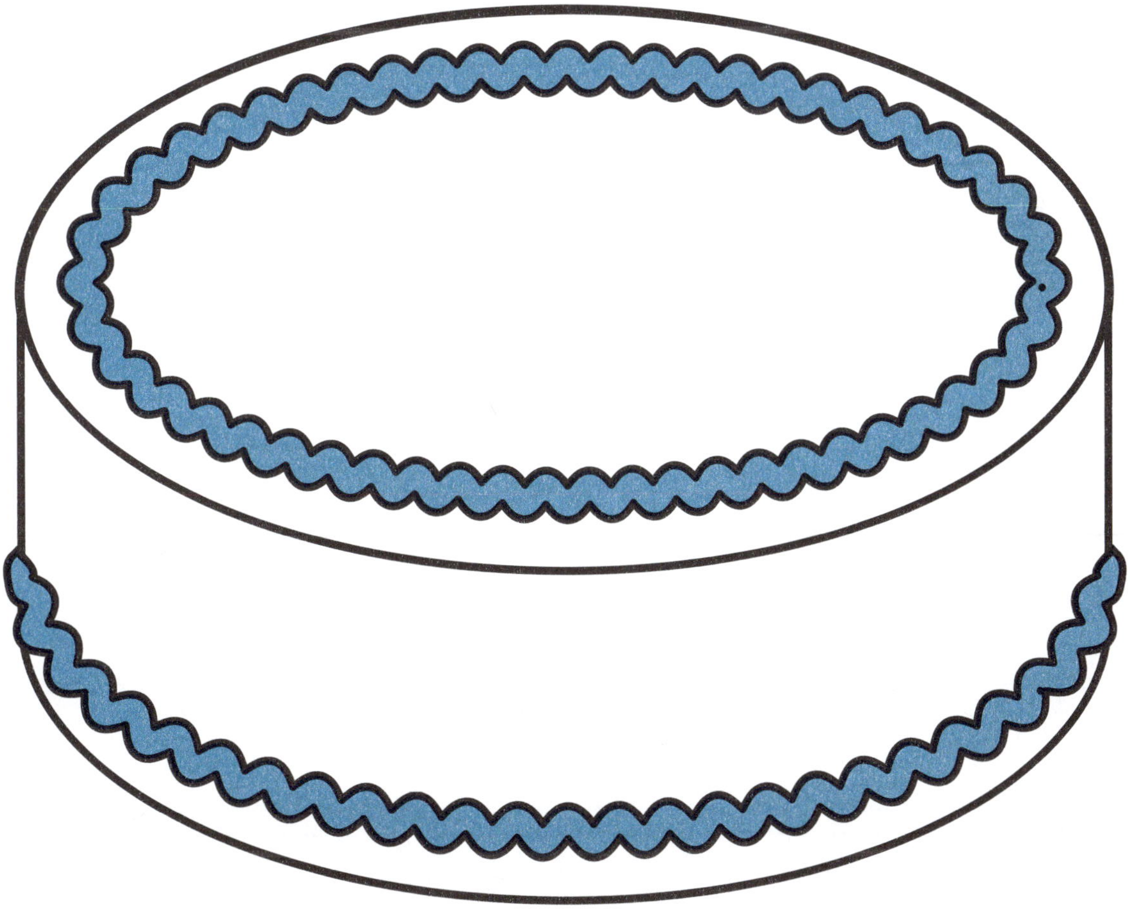

Number Path

REMOVABLE

1	2	3	4	5	6	7	8	9	10

Name ______________________________

PROBLEM SET

I can show how addition and subtraction are alike with a number bond. **I can** show how they are different with a number sentence.

California CCSS-M
K.OA.A.1, K.OA.A.2

Tell a story to match the picture. Fill in the number bond and number sentence.

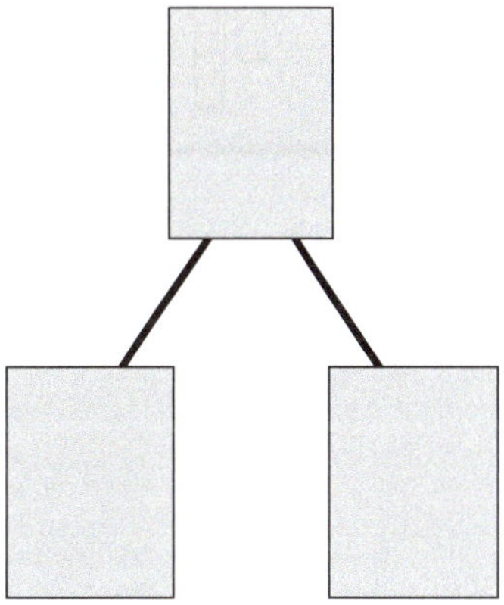

4 + 2 = ☐

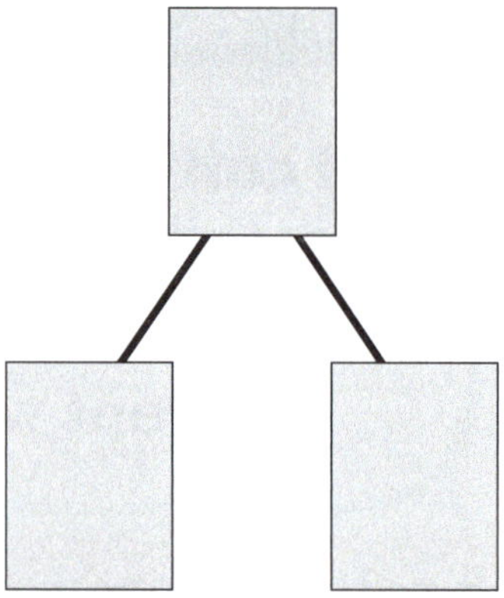

6 − ☐ = ☐

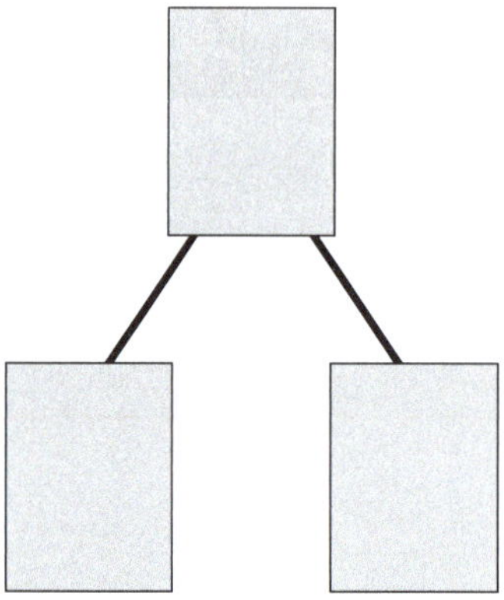

Fill in the number sentences.

$2 + 2 = \square$

$2 + 3 = \square$

$2 + 4 = \square$

$4 - 2 = \square$

$5 - 2 = \square$

$6 - 2 = \square$

Scoreboard

1	2	3	4	5	6	7	8	9	10

Name ______________________________

PROBLEM SET

I can start counting from different numbers to find the total.

California CCSS-M
K.CC.A.2

Count. Write how many.

4

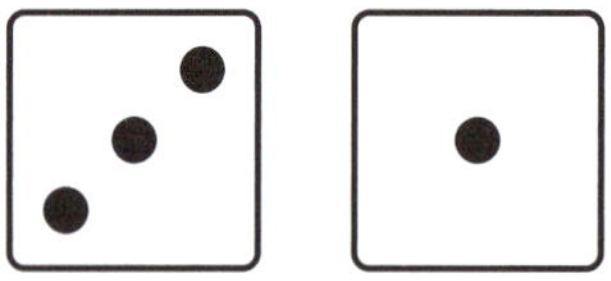

Name ______________________________

LESSON

I can choose tools to find partners to 10 with weight and length. **I can** write a number sentence to match.

California CCSS-M
K.OA.A.4

Fill in the number sentence.

☐ + ☐ = 10

1	2	3	4	5	6	7	8	9	10

☐ + ☐ = 10

1	2	3	4	5	6	7	8	9	10

Name ______________________________

PROBLEM SET

I can draw to show the parts and total. **I can** write a number sentence to match.

California CCSS-M
K.OA.A.1

Draw more dots to make 10. Fill in the number bond.

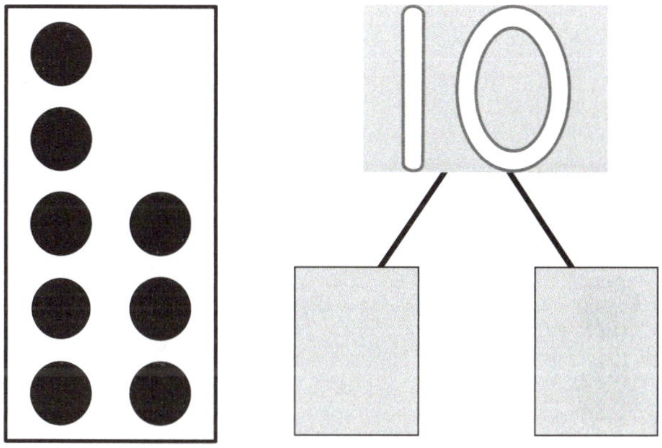

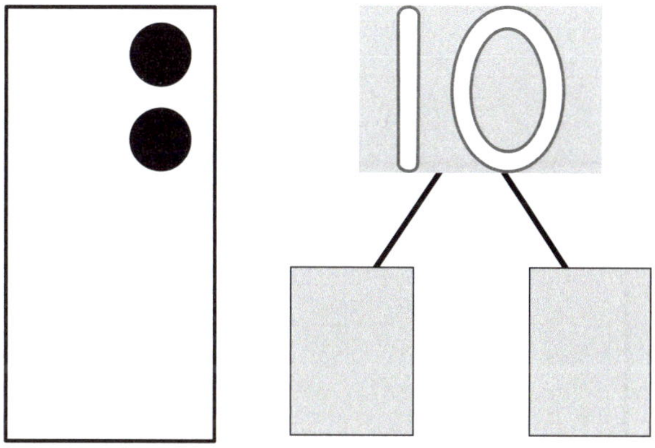

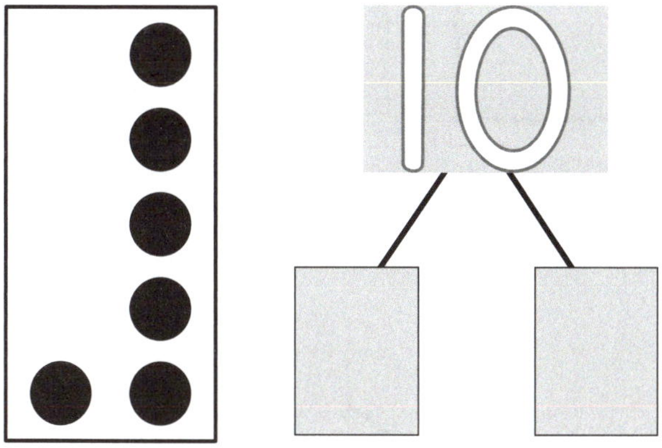

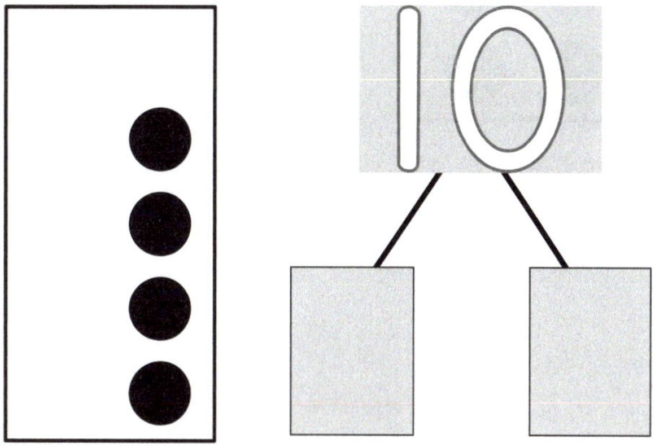

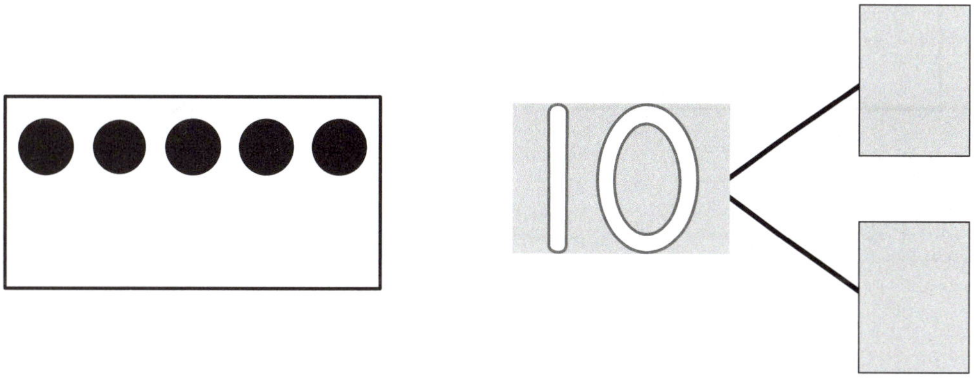

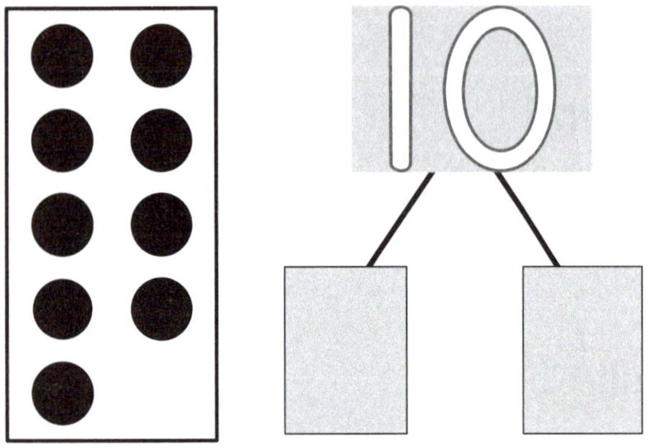
10

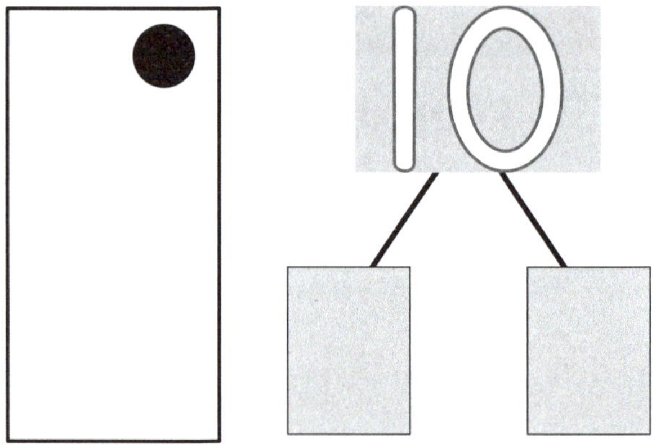
10

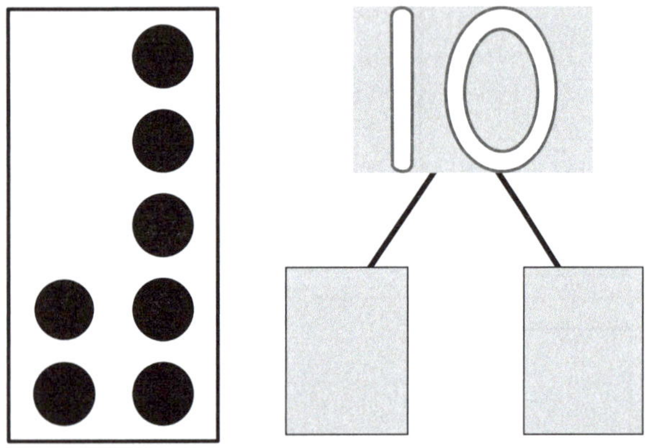
10

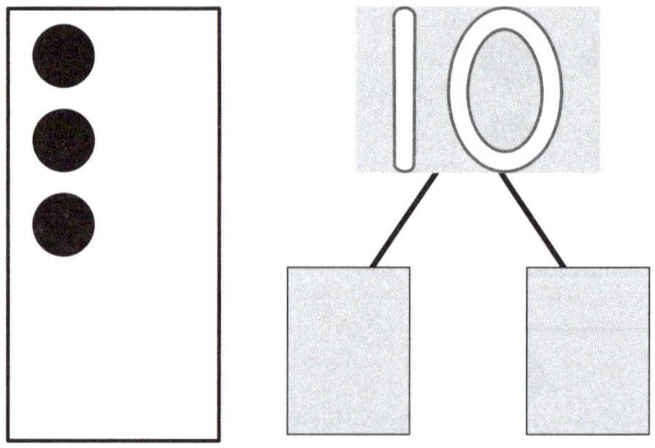
10

10

MODULE 5
TOPIC D

FAMILY MATH

Make Use of Structure

Key Term

Pattern

Dear Family,

Students build an understanding of structure by studying and naming patterns. For example, green- orange, green-orange can be described as an AB pattern. Students demonstrate their understanding by recreating the same type of pattern with different materials. When analyzing patterns, students are likely to perceive them in different ways. This provides an important opportunity to share multiple perspectives, a theme that will carry throughout their mathematics learning.

This green-orange pattern is an example of an AB pattern.

At-Home Activities

Activity Idea 1 Pattern Hunt

Invite your child to help you hunt for patterns. Consider patterns that are in and outside of your home. Children may notice patterns in colors, shapes, or design. As your child points out a pattern, pause to ask them how they see it. Consider highlighting that people see things differently and acknowledge their viewpoint.

Activity Idea 2 Match My Pattern

Build a pattern using household items, such as dried beans and pasta. It might look like this: bean-bean-pasta, bean-bean-pasta. Ask your child to study the pattern and to use other items to show the same pattern. They might recreate the pattern using coins, such as, penny-penny-dime, penny-penny-dime.

Patterns

Students learn that a pattern is something that repeats and can help us know what will happen next. In the example below, students might see a color or size pattern (i.e., big-small, big-small or red-red-blue, red-red-blue).

As students replicate a pattern with different items, they discover mathematics is built on structure, not physical appearances. As their learning progresses, students can use repeated reasoning to recognize patterns in a story context. For example, by solving a series of problems, students might notice, *When there are 5 children in the bouncy house, there are 10 shoes outside. When there are 6 children, there are 12 shoes. And when there are 7 children, there are 14 shoes.* The pattern can be used to efficiently solve the next problem, *How many shoes will be outside the bouncy house when there are 8 children?*

Children	Shoes
5	10
6	12
7	14

Students may notice that when 1 more child goes into the bouncy house, the number of shoes goes up by 2 (or the number doubles). 5 and 5 is 10. 6 and 6 is 12. Students can use these patterns to solve the next problem (8 children in the bouncy house) by either adding 2 to the number of shoes or by doubling 8.

Contemplate Then Calculate

5-Groups | FLUENCY SHEET 1

Write how many dots.

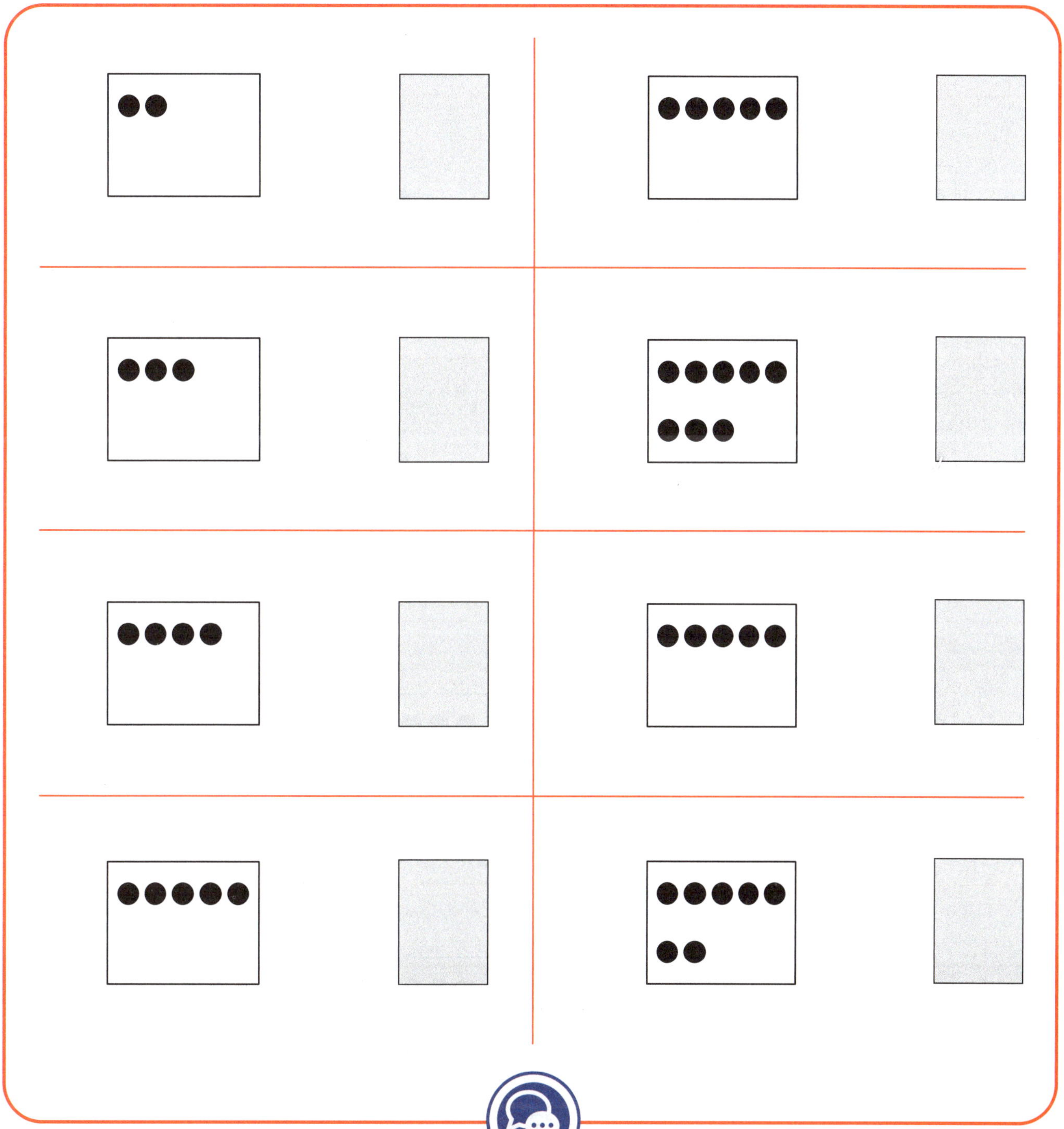

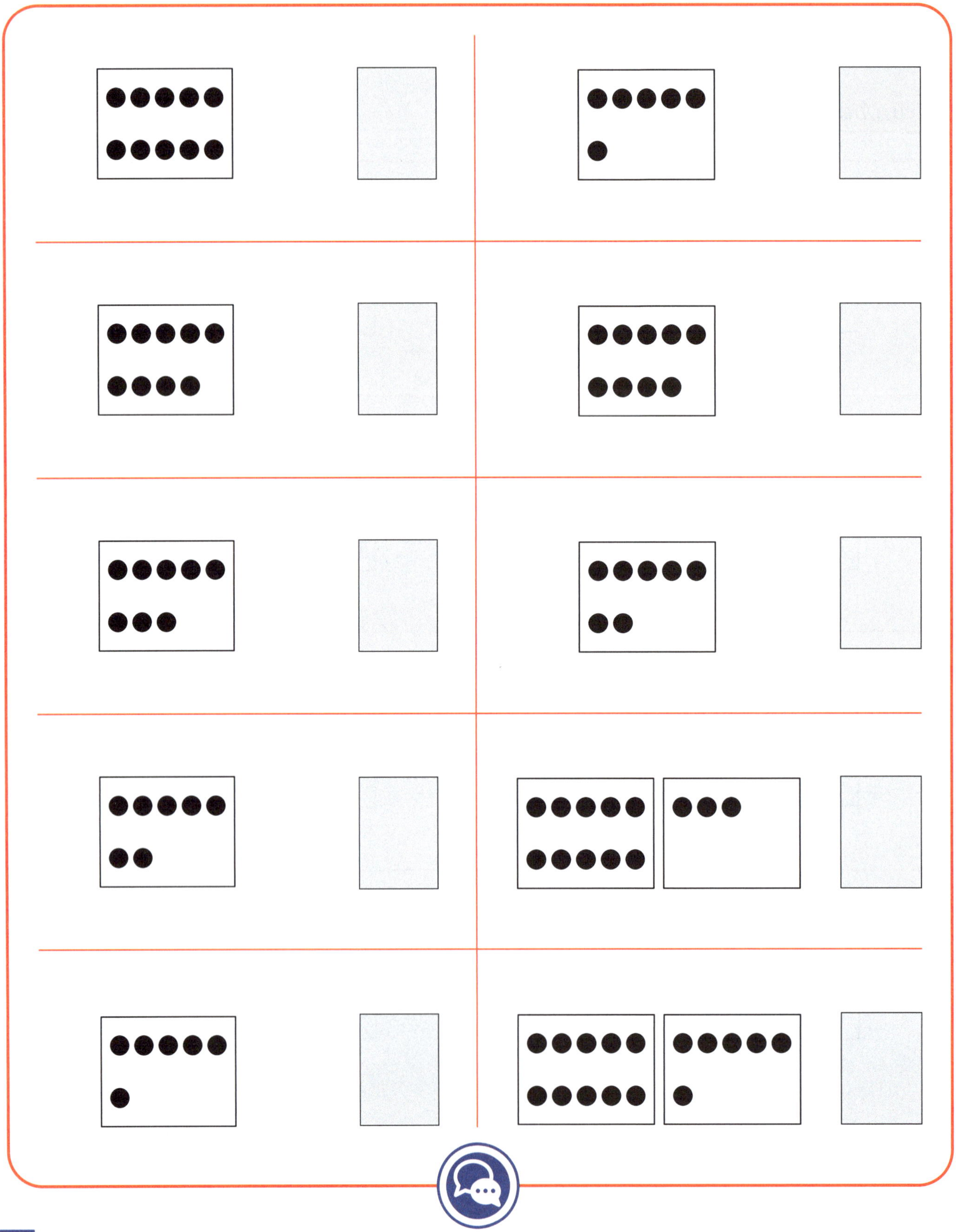

Contemplate Then Calculate

5-Groups | FLUENCY SHEET 2

Write how many dots.

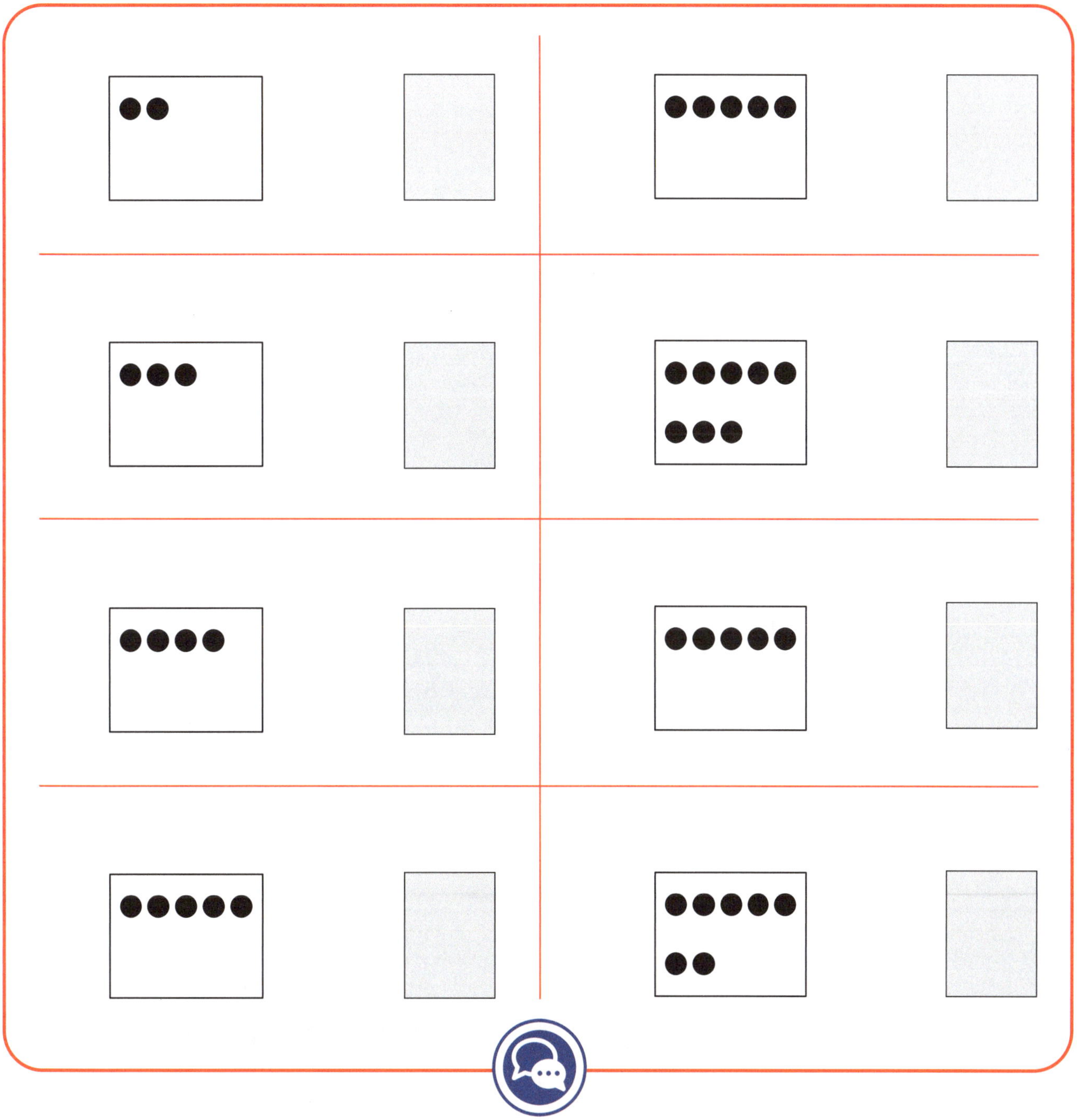

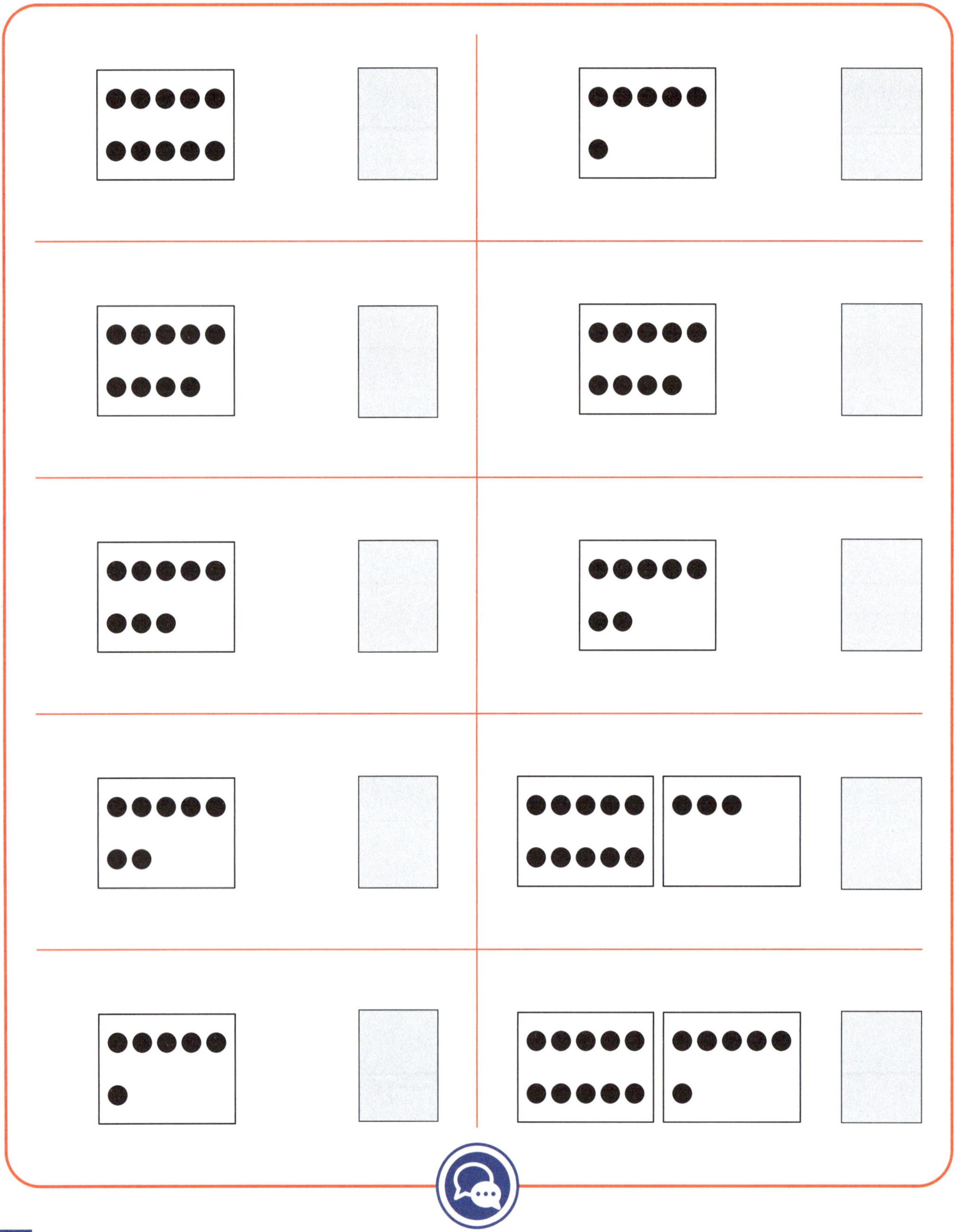

Name ____________________

PROBLEM SET

I can find a pattern. **I can** add more to a pattern.

California CCSS-M
K.CC.A.2

Build the pattern with cubes. Finish the pattern.

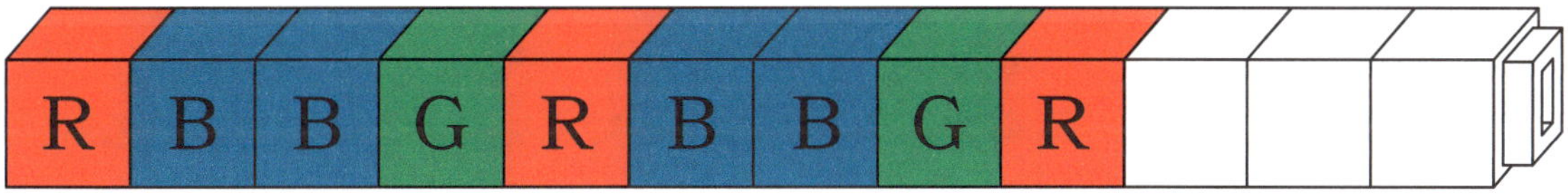

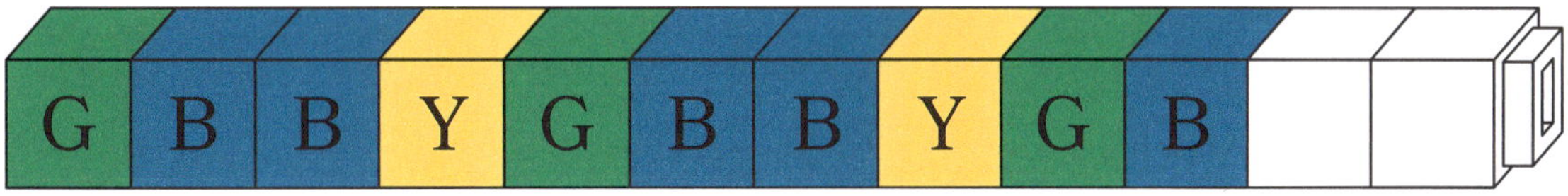

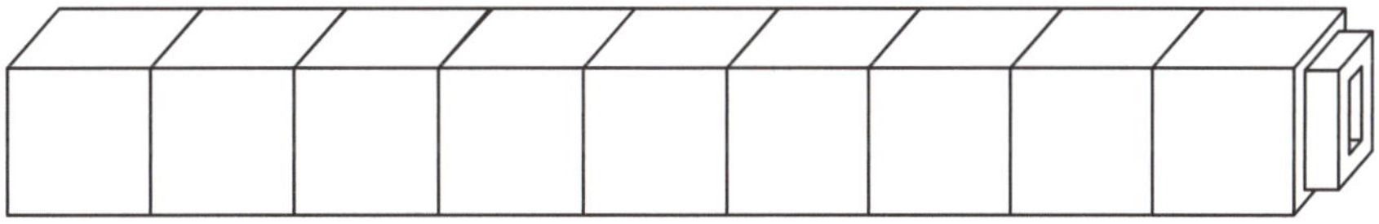

Recording Sheet

REMOVABLE

1	2	3	4	5	6	7	8	9	10	11	12	13	14	15

Circle Groups of 2

REMOVABLE

Circle a group of 2 in each picture.

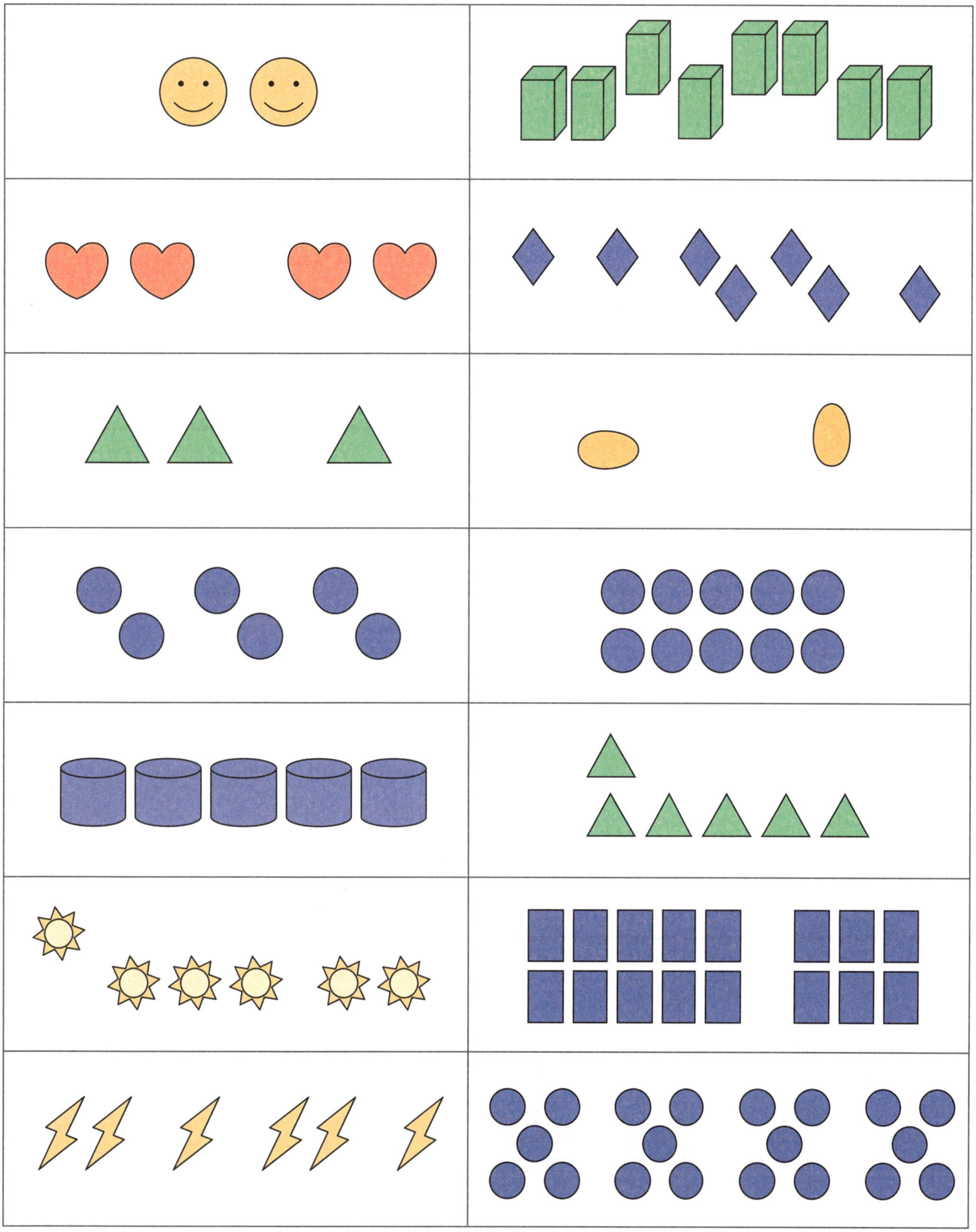

Name ______________________________

PROBLEM SET

I can use a pattern to solve a story problem. **I can** find groups of 2 and write a number sentence to match.

California CCSS-M
K.OA.A.1

Circle groups of 2. Write the number sentence.

☐ + ☐ = ☐

Circle groups of 2. Write the number sentence.

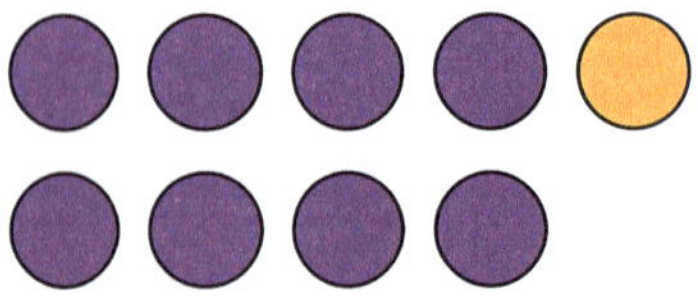

Circle Groups of 3

REMOVABLE

Circle a group of 3 in each picture.

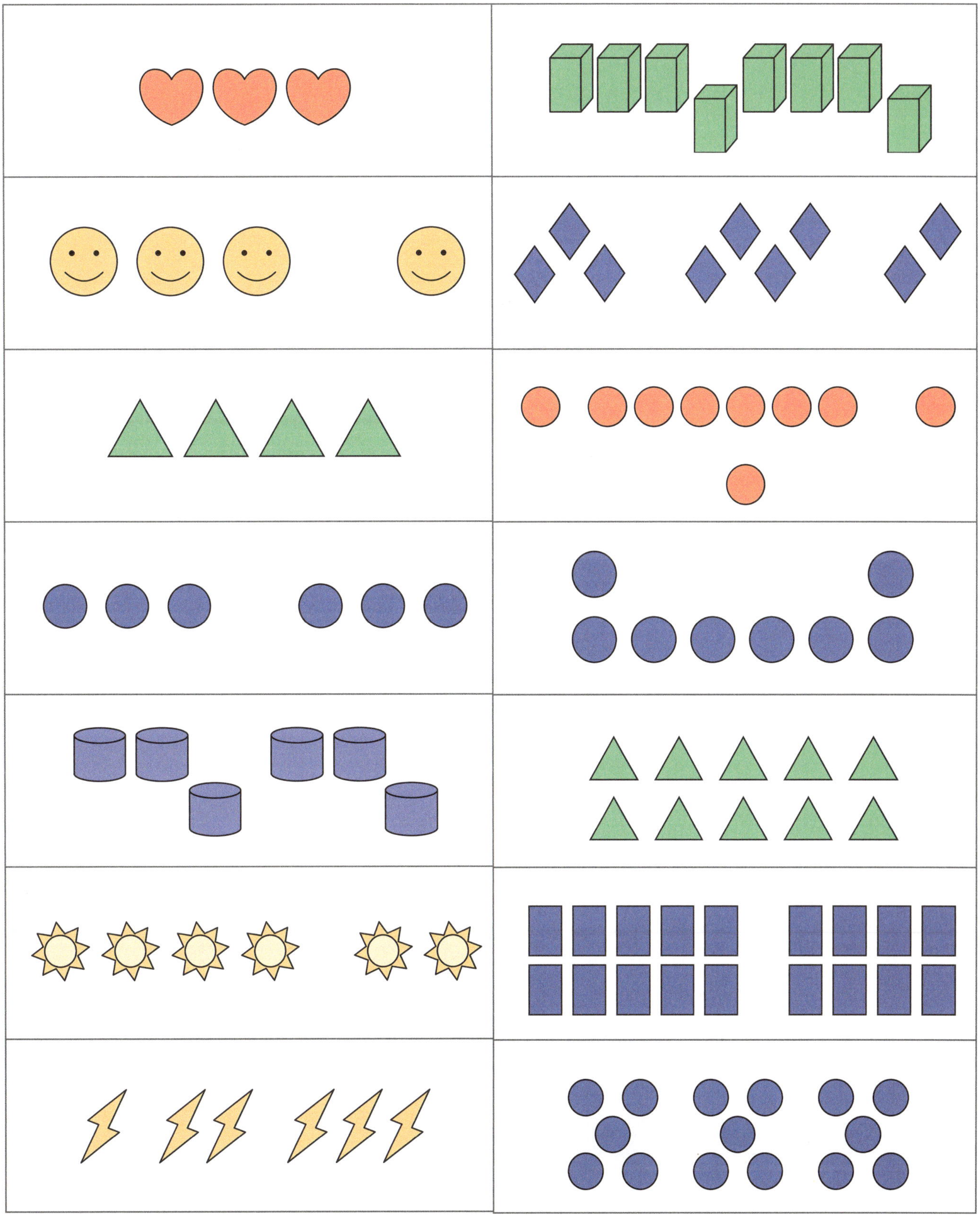

Quilt

REMOVABLE

Number Sentence Recording Sheet

REMOVABLE

1	2	3	4	5	6	7	8	9	10

	+		=	5

	−		=	4

1 0	=		+	

	−		=	7

California CCSS-M
K.OA.A.1, K.OA.A.4

Name ____________________

LESSON

I can find patterns to solve problems. **I can** use numbers to make true number sentences. **I can** find different partners to 10.

Fill in the pattern.

In	Out	Pattern
		+ ☐
2 →	4	+ ☐
3 →	5	+ ☐

In	Out	Pattern
		+ ☐
4 →	6	+ ☐
5 →	7	+ ☐

In	Out	Pattern
		+
3 →		+ 3
4 →		+ 3
5 →		+ 3
6 →		+ 3

California CCSS-M
Supports K.CC.1–5

Name ______________________

LESSON

I can use groups of 5 or 10 to count a collection. **I can** write a number sentence to show how I counted.

Recording Sheet

Draw your collection to show how you counted.

Name ______________________________

Trace and write the numeral 0.

Trace a hexagon. Count the corners.

corners

Count and write how many. Circle the one with fewer.

Name ______________________

Circle the one that is LONGER.

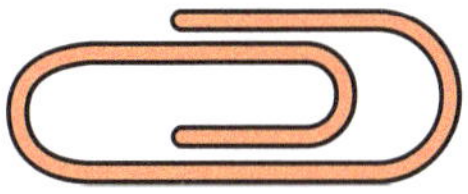

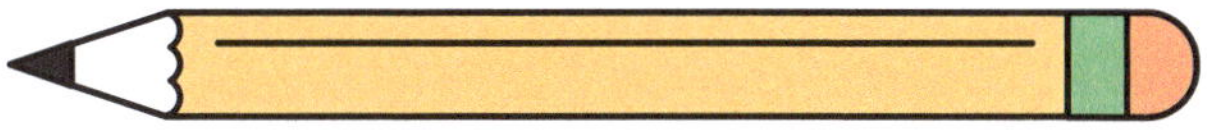

Fill in the missing number.

0 1 ☐ 3 4 5	5 ☐ 3 2 1 0
0 1 2 ☐ 4 5	☐ 1 2 3 4 5

Circle the parts. Fill in the number sentence.

☐ is ☐ and ☐ .

☐ is ☐ and ☐ .

Read Aloud

Terminology

addition

When we put parts together to find the total, we call that **addition**.

(Lesson 1)

$3 + 2 = 5$

equals

Equals means “is” or “make.”

(Lesson 1)

9 is 7 and 2.

9 = 7 + 2

↑

7 and 2 make 9.

7 + 2 = 9

↑

minus

When subtracting, we say **minus** or use a **minus** sign instead of “take away.”

(Lesson 9)

9 take away 4 is 5

9 - 4 = 5

↑

pattern

When we see a **pattern,** we know what comes next.

(Lesson 22)

B s B s B s B s

plus

When adding, we say **plus** or use a **plus** sign instead of “and.”

(Lesson 1)

7 and 2 make 9.

7 + 2 = 9

subtraction

When something gets taken away and we find how many are left, we call it **subtraction**.

(Lesson 8)

$4 - 1 = 3$

Standards

Module Content Standards

Know number names and the count sequence.

K.CC.A.2 Count forward beginning from a given number within the known sequence (instead of having to begin at 1).

Understand addition as putting together and adding to, and understand subtraction as taking apart and taking from.

K.OA.A.1 Represent addition and subtraction with objects, fingers, mental images, drawings[2], sounds (e.g., claps), acting out situations, verbal explanations, expressions, or equations.

K.OA.A.2 Solve addition and subtraction word problems, and add and subtract within 10, e.g., by using objects or drawings to represent the problem.

K.OA.A.3 Decompose numbers less than or equal to 10 into pairs in more than one way, e.g., by using objects or drawings, and record each decomposition by a drawing or equation (e.g., 5 = 2 + 3 and 5 = 4 + 1).

K.OA.A.4 For any number from 1 to 9, find the number that makes 10 when added to the given number, e.g., by using objects or drawings, and record the answer with a drawing or equation.

K.OA.A.5 Fluently add and subtract within 5.

Analyze, compare, create, and compose shapes.

K.G.B.6 Compose simple shapes to form larger shapes. *For example, "Can you join these two triangles with full sides touching to make a rectangle?"*

2 Drawings need not show details, but should show the mathematics in the problem.

Standards for Mathematical Practice

SMP.1 Make sense of problems and persevere in solving them.

SMP.2 Reason abstractly and quantitatively.

SMP.3 Construct viable arguments and critique the reasoning of others.

SMP.4 Model with mathematics.

SMP.5 Use appropriate tools strategically.

SMP.6 Attend to precision.

SMP.7 Look for and make use of structure.

SMP.8 Look for and express regularity in repeated reasoning.

Alignment Overview

This chart identifies the content standards, Standards for Mathematical Practice (SMPs), and Big Ideas for each lesson in the module.

Lesson	Standards	SMPs	Big Ideas
1	K.OA.A.1	2	Being flexible within 10 Model with Numbers
2	K.OA.A.1	7	Being flexible within 10 Model with Numbers
3	K.OA.A.1 K.OA.A.2	5	Being flexible within 10 Model with Numbers
4	K.OA.A.1 K.OA.A.3	6	Being flexible within 10 Model with Numbers
5	K.OA.A.1	2	Being flexible within 10 Model with Numbers
6	K.OA.A.1	3	Being flexible within 10 Model with Numbers
7	K.OA.A.1 K.OA.A.5	5	Being flexible within 10 Model with Numbers
8	K.OA.A.1	8	Being flexible within 10 Model with Numbers
9	K.OA.A.1	2	Being flexible within 10 Model with Numbers
10	K.OA.A.1 K.OA.A.2	5	Being flexible within 10 Model with Numbers
11	K.OA.A.1	7	Being flexible within 10 Model with Numbers
12	K.OA.A.1 K.OA.A.2	4	Being flexible within 10 Model with Numbers
13	K.OA.A.1	3	Being flexible within 10 Model with Numbers
14	K.OA.A.1 K.OA.A.5	5	Being flexible within 10 Model with Numbers

Lesson	Standards	SMPs	Big Ideas
15	K.OA.A.1 K.OA.A.2	1	Being flexible within 10 Model with Numbers
16	K.OA.A.1 K.OA.A.2	7	Being flexible within 10 Model with Numbers
17	K.OA.A.2	1	Being flexible within 10 Model with Numbers
18	K.CC.A.2	8	How Many?
19	K.OA.A.1	4	Being flexible within 10 Model with Numbers
20	K.OA.A.4	3	Being flexible within 10
21	K.OA.A.1	2 and 4	Being flexible within 10 Model with Numbers
22	K.CC.A.2	7	How Many?
23	K.CC.A.2	2, 7, 8	How Many?
24	K.OA.A.1	7 and 8	Being flexible within 10 Model with Numbers
25	K.G.B.6	7	Being flexible within 10
26	K.OA.A.1 K.OA.A.4	8	Being flexible within 10 Model with Numbers
27	Supports K.CC.1–5	7	How Many?

Bibliography

California Department of Education. *California Common Core State Standards: Mathematics.* Sacramento, CA: California Department of Education, 2014.

California Department of Education. *Mathematics Framework for California Public Schools: Kindergarten Through Grade Twelve*. Sacramento, CA: California Department of Education, 2023.

Credits

Great Minds has made every effort to obtain permission for the reprinting of all copyrighted material. If any owner of copyrighted material is not acknowledged herein, please contact Great Minds for proper acknowledgment in all future editions and reprints of this module.

Cover, Adam Burton/Alamy Stock Photo: page 113, Steve Skjold/Alamy Stock Photo; All other images are the property of Great Minds.

For a complete list of credits, visit http://eurmath.link/media-credits.

Acknowledgments

Kate Austin, Beth Barnes, Melissa Brown, Dawn Burns, Stella Chen, Mary Christensen-Cooper, Dr. Hazel Coltharp, Nicole Conforti, Cheri DeBusk, Stephanie DeGiulio, Jill Diniz, Brittany duPont, Lacy Endo-Peery, Krysta Gibbs, January Gordon, Melanie Gutierrez, Eddie Hampton, Brittney Hawkins, Tiffany Hill, Rachel Hylton, Travis Jones, Kelly Kagamas Tomkies, Julie Katzen, Emily Koesters, Liz Krisher, Ben McCarty, Kate McGill Austin, Maureen McNamara Jones, Marie Libassi-Behr, Bobbe Maier, Cristina Metcalf, Ashley Meyer, Melissa Mink, Katie Moore, Dave Morris, Bruce Myers, Marya Myers, Lauren Nelson, Andrea Neophytou Hart, Geoff Patterson, Maximilian Peiler-Burrows, Shelley Petre, John Reynolds, Meri Robie-Craven, Colleen Sheeron-Laurie, Erika Silva, Jessica Sims, Robyn Sorenson, Paul Spicer, Lori Sponenburgh, Tara Stewart, Julie Stoehr, James Tanton, Julia Tessler, Janel Verrilli, Philippa Walker, Rachael Waltke, Jamie Wells Schneider

Ana Alvarez, Lynne Askin-Roush, Stephanie Bandrowsky, Mariel Bard, Rebeca Barroso, Brianna Bemel, Rebecca Blaho, Charles Blake, Carolyn Buck, Lisa Buckley, Shanice Burton, Adam Cardais, Cindy Carlone, Gina Castillo, Ming Chan, Tatyana Chapin, Christina Cooper, Kim Cotter, Gary Crespo, Lisa Crowe, David Cummings, Brandon Dawley, Cherry dela Victoria, Timothy Delaney, Delsena Draper, Erin DuRant, Sandy Engelman, Tamara Estrada, Ubaldo Feliciano-Hernández, Soudea Forbes, Liz Gabbard, Diana Ghazzawi, Lisa Giddens-White, Laurie Gonsoulin, Adam Green, Sagal Hassan, Kristen Hayes, Tim Heppner, Marcela Hernandez, Sary Hernandez, Abbi Hoerst, Elizabeth Jacobsen, Ashley Kelley, Sonia Khaleel, Lisa King, Sarah Kopec, Drew Krepp, Jenny Loomis, Stephanie Maldonado, Christina Martire, Siena Mazero, Thomas McNeely, Cindy Medici, Ivonne Mercado, Sandra Mercado, Brian Methe, Sara Miller, Mary-Lise Nazaire, Corinne Newbegin, Tara O'Hare, Max Oosterbaan, Tamara Otto, Christine Palmtag, Laura Parker, Toy Parrish, Katie Prince, Neha Priya, Jeff Robinson, Nate Robinson, Gilbert Rodriguez, Todd Rogers, Karen Rollhauser, Neela Roy, Gina Schenck, Aaron Shields, Madhu Singh, Leigh Sterten, Mary Sudul, Lisa Sweeney, Tracy Vigliotti, Bruce Vogel, Charmaine Whitman, Glenda Wisenburn-Burke, Samantha Wofford, Howard Yaffe, Dani Zamora

TALKING TOOL

I Can Share My Thinking	My drawing shows . . . . I did it this way because . . . . I think ___ because . . . .
I Can Agree or Disagree	I agree because . . . . I disagree because . . . . I did it a different way. I . . . .
I Can Ask Questions	How did you . . . ? Why did you . . . ? Can you explain . . . ?
I Can Say It Again	I heard you say . . . . ___ said . . . . Can you say it another way?